ZIML Math Competition Book

Junior Varsity 2018-2019

Areteem Institute

ZIML Math Competition Book Junior Varsity 2018-2019

Edited by   John Lensmire
             David Reynoso
             Kelly Ren
             Kevin Wang

Copyright © 2019 ARETEEM INSTITUTE
WWW.ARETEEM.ORG

PUBLISHED BY ARETEEM PRESS

ISBN-10: 1-944863-47-8
ISBN-13: 978-1-944863-47-0

First printing, August 2019.

TITLES PUBLISHED BY ARETEEM PRESS

Cracking the High School Math Competitions (and Solutions Manual) - Covering AMC 10 & 12, ARML, and ZIML
Mathematical Wisdom in Everyday Life (and Solutions Manual) - From Common Core to Math Competitions
Geometry Problem Solving for Middle School (and Solutions Manual) - From Common Core to Math Competitions
Fun Math Problem Solving For Elementary School (and Solutions Manual)

ZIML MATH COMPETITION BOOK SERIES

ZIML Math Competition Book Division E 2016-2017
ZIML Math Competition Book Division M 2016-2017
ZIML Math Competition Book Division H 2016-2017
ZIML Math Competition Book Jr Varsity 2016-2017
ZIML Math Competition Book Varsity Division 2016-2017
ZIML Math Competition Book Division E 2017-2018
ZIML Math Competition Book Division M 2017-2018
ZIML Math Competition Book Division H 2017-2018
ZIML Math Competition Book Jr Varsity 2017-2018
ZIML Math Competition Book Varsity Division 2017-2018
ZIML Math Competition Book Division E 2018-2019
ZIML Math Competition Book Division M 2018-2019
ZIML Math Competition Book Division H 2018-2019
ZIML Math Competition Book Jr Varsity 2018-2019
ZIML Math Competition Book Varsity Division 2018-2019

MATH CHALLENGE CURRICULUM TEXTBOOKS SERIES

Math Challenge I-A Pre-Algebra and Word Problems
Math Challenge I-B Pre-Algebra and Word Problems
Math Challenge I-C Algebra
Math Challenge II-A Algebra
Math Challenge II-B Algebra
Math Challenge III Algebra

Math Challenge I-A Geometry
Math Challenge I-B Geometry
Math Challenge I-C Topics in Algebra
Math Challenge II-A Geometry
Math Challenge II-B Geometry
Math Challenge III Geometry
Math Challenge I-A Counting and Probability
Math Challenge I-B Counting and Probability
Math Challenge I-C Geometry
Math Challenge II-A Combinatorics
Math Challenge II-B Combinatorics
Math Challenge III Combinatorics
Math Challenge I-A Number Theory
Math Challenge I-B Number Theory
Math Challenge I-C Finite Math
Math Challenge II-A Number Theory
Math Challenge II-B Number Theory
Math Challenge III Number Theory

COMING SOON FROM ARETEEM PRESS

Fun Math Problem Solving For Elementary School Vol. 2 (and Solutions Manual)
Counting & Probability for Middle School (and Solutions Manual) - From Common Core to Math Competitions
Number Theory Problem Solving for Middle School (and Solutions Manual) - From Common Core to Math Competitions

The books are available in paperback and eBook formats (including Kindle and other formats).
To order the books, visit https://areteem.org/bookstore.

Contents

Introduction

Each month during the school year, Areteem Institute hosts the online Zoom International Math League (ZIML) competitions. Students can compete in one of five divisions based on their age and mathematical level (details shown on Page 9).

This book contains the problems, answers, and full solutions from the nine ZIML Junior Varsity Competitions held during the 2018-2019 School Year. It is divided into three parts:

1. The complete Junior Varsity ZIML Competitions (20 questions per competition) from October 2018 to June 2019.
2. The solutions for each of the competitions, including detailed work and helpful tricks.
3. An appendix including the topics and knowledge points covered for Junior Varsity, a glossary including common mathematical terms, and answer keys for each of the competitions so students can easily check their work.

The questions found on the ZIML competitions are meant to test your problem solving skills and train you to apply the knowledge you know to many different applications. We hope you enjoy the problems!

About Zoom International Math League

The Zoom International Math League (ZIML) has a simple goal: provide a platform for students to build and share their passion for math and other STEM fields with students from around the globe. Started in 2008 as the Southern California Mathematical Olympiad, ZIML has a rich history of past participants who have advanced to top tier colleges and prestigious math competitions, including American Math Competitions, MATHCOUNTS, and the International Math Olympaid.

The ZIML Core Online Programs, most available with a free account at `ziml.areteem.org`, include:

- **Daily Magic Spells:** Provides a problem a day (Monday through Friday) for students to practice, with full solutions available the next day.
- **Weekly Brain Potions:** Provides one problem per week posted in the online discussion forum at `ziml.areteem.org`. Usually the problem does not have a simple answer, and students can join the discussion to share their thoughts regarding the scenarios described in the problem, explore the math concepts behind the problem, give solutions, and also ask further questions.
- **Monthly Contests:** The ZIML Monthly Contests are held the first weekend of each month during the school year (October through June). Students can compete in one of 5 divisions to test their knowledge and determine their strengths and weaknesses, with winners announced after the competition.
- **Math Competition Practice:** The Practice page contains sample ZIML contests and an archive of AMC-series tests for online practice. The practices simulate the real contest environment with time-limits of the contests automatically controlled by the server.
- **Online Discussion Forum:** The Online Discussion Forum

is open for any comments and questions. Other discussions, such as hard Daily Magic Spells or the Weekly Brain Potions are also posted here.

These programs encourage students to participate consistently, so they can track their progress and improvement each year.

In addition to the online programs, ZIML also hosts onsite Local Tournaments and Workshops in various locations in the United States. Each summer, there are onsite ZIML Competitions at held at Areteem Summer Programs, including the International ZIML Convention, which is a two day convention with one day of workshops and one day of competition.

ZIML Monthly Contests are organized into five divisions ranging from upper elementary school to advanced material based on high school math.

- **Varsity:** This is the top division. It covers material on the level of the last 10 questions on the AMC 12 and AIME level. This division is open to all age levels.
- **Junior Varsity:** This is the second highest competition division. It covers material at the AMC 10/12 level and State and National MathCounts level. This division is open to all age levels.
- **Division H:** This division focuses on material from a standard high school curriculum. It covers topics up to and including pre-calculus. This division will serve as excellent practice for students preparing for the math portions of the SAT or ACT. This division is open to all age levels.
- **Division M:** This division focuses on problem solving using math concepts from a standard middle school math curriculum. It covers material at the level of AMC 8 and School or Chapter MathCounts. This division is open to all students who have not started grade 9.

- **Division E:** This division focuses on advanced problem solving with mathematical concepts from upper elementary school. It covers material at a level comparable to MOEMS Division E. This division is open to all students who have not started grade 6.

This problem book features the Junior Varsity Contests. For a detailed list of topics covered for Junior Varsity see p.203 in the Appendix.

To participate in the ZIML Online Programs, create a free account at ziml.areteem.org. The ZIML site features are also provided on the ZIML Mobile App, which is available for download from Apple's App Store and Google Play Store.

About Areteem Institute

Areteem Institute is an educational institution that develops and provides in-depth and advanced math and science programs for K-12 (Elementary School, Middle School, and High School) students and teachers. Areteem programs are accredited supplementary programs by the Western Association of Schools and Colleges (WASC). Students may attend the Areteem Institute in one or more of the following options:

- Live and real-time face-to-face online classes with audio, video, interactive online whiteboard, and text chatting capabilities;
- Self-paced classes by watching the recordings of the live classes;
- Short video courses for trending math, science, technology, engineering, English, and social studies topics;
- Summer Intensive Camps held on prestigious university campuses and Winter Boot Camps;
- Practice with selected free daily problems and monthly ZIML competitions at `ziml.areteem.org`.

Areteem courses are designed and developed by educational experts and industry professionals to bring real world applications into STEM education. The programs are ideal for students who wish to build their mathematical strength in order to excel academically and eventually win in Math Competitions (AMC, AIME, USAMO, IMO, ARML, MathCounts, Math Olympiad, ZIML, and other math leagues and tournaments, etc.), Science Fairs (County Science Fairs, State Science Fairs, national programs like Intel Science and Engineering Fair, etc.) and Science Olympiads, or for students who purely want to enrich their academic lives by taking more challenging courses and developing outstanding analytical, logical, and creative problem solving skills.

Since 2004 Areteem Institute has been teaching with methodology that is highly promoted by the new Common Core State Standards: stressing the conceptual level understanding of the math concepts, problem solving techniques, and solving problems with real world applications. With the guidance from experienced and passionate professors, students are motivated to explore concepts deeper by identifying an interesting problem, researching it, analyzing it, and using a critical thinking approach to come up with multiple solutions.

Thousands of math students who have been trained at Areteem have achieved top honors and earned top awards in major national and international math competitions, including Gold Medalists in the International Math Olympiad (IMO), top winners and qualifiers at the USA Math Olympiad (USAMO/JMO) and AIME, top winners at the Zoom International Math League (ZIML), and top winners at the MathCounts National Competition. Many Areteem Alumni have graduated from high school and gone on to enter their dream colleges such as MIT, Cal Tech, Harvard, Stanford, Yale, Princeton, U Penn, Harvey Mudd College, UC Berkeley, or UCLA. Those who have graduated from colleges are now playing important roles in their fields of endeavor.

Further information about Areteem Institute, as well as updates and errata of this book, can be found online at http://www. areteem.org.

Acknowledgments

This book contains the Online ZIML Junior Varsity Problems from the 2018-19 school year. These problems were created and compiled by the staff of Areteem Institute. These problems were inspired by questions from the Areteem Math Challenge Courses, past questions on the ACT/SAT/GRE, past math competitions, math textbooks, and countless other resources and people encountered by the Areteem Curriculum Department in their life devoted to math. We thank all these sources for growing and nurturing our passion for math.

The Areteem staff, including John Lensmire, David Reynoso, Kevin Wang, and Kelly Ren, are the main contributors who compiled, edited, and reviewed this book.

Lastly, thanks to all the students who have participated and continue to participate in the Zoom International Math League. Your dedication to the Daily Magic Spells and Monthly Contests makes all of this possible, and we hope you continue to enjoy ZIML for years to come!

1. ZIML Contests

This part of the book contains the Junior Varsity ZIML Contests from the 2018-19 School Year. There were nine monthly competitions, held on the dates found below:

- October 5-7
- November 2-4
- December 7-9
- January 4-6
- February 1-3
- March 1-3
- April 5-7
- May 3-5
- June 7-9

1.1 ZIML October 2018 Junior Varsity

Below are the 20 Problems from the Junior Varsity ZIML Competition held in October 2018.
The answer key is available on p.216 in the Appendix.
Full solutions to these questions are available starting on p.84.

Problem 1
If $(x^5 - x^4 - x^3 - x - 2)(x^4 + 2x^3 + 4x^2 + 2x + 1)$ is expanded, what is the coefficient of x^5?

Problem 2
3 freshman, 3 sophomores, 3 juniors, and 3 seniors were awarded medals for their community service work over the summer. They all attended an award dinner and sat, in a line, so that all the freshmen were together, all the sophomores were together, etc., but not in any particular order otherwise. How many different seating arrangements were possible?

Problem 3
The six-digit number $\overline{1A525B}$ is divisible by 308. What is this number?

Problem 4

George was learning about polynomials and (after some work) found the 5 real solutions r, s, t, u, v to the quintic polynomial

$$3x^5 + 11x^4 - 30x^3 - 117x^2 - 107x - 30 = 0.$$

To explore the relationship between the roots, he decided to make a multiplication chart with the roots:

$\times$	r	s	t	u	v
r	r^2	rs	rt	$\cdots$	
s	sr	s^2	$\cdots$		
t	tr	$\cdots$			
u	$\cdots$				
v					

George added up all 25 elements in his table and got a fraction $\dfrac{N}{M}$ for positive integers N, M with $\gcd(N, M) = 1$. What is $N + M$?

Problem 5

Find the smallest integer K such that (i) K is a multiple of 66, (ii) K has at least one pair of repeated digits, and (iii) K has an odd number of factors. What is K?

Problem 6

A convex polygon with n sides has interior angles of

$$30°, 60°, 90°, \ldots, n \cdot 30°.$$

What is the sum of all possible n?

Problem 7

At the end of the day, teacher Mr. David wrote the 6 numbers he wanted to remember for the next day on the board: 6, 8, 2, 3, and 9. Unfortunately one of his mischievous students took each number one by one, raised it to the Nth power, and replaced it with the remainder when the number was divided by 13 on the board.

The next day, Mr. David noticed and warned his class not to do this again. He again wrote the six number (same as before) on the board. Not listening, the mischievous student again took each number one by one, this time raised it to the $(N+5)$th power, and replaced it with the remainder when divided by 13 on the board. To his surprise, this time none of the numbers changed!

What is the smallest 3-digit integer N for which this is possible?

Problem 8

Dominic is building an unfair coin, where the probability of getting heads $P(H)$ is greater than getting tails $P(T)$. To avoid suspicion, he wants to make $P(H)$ as low as possible, yet he wants to ensure that the probability of getting at least one heads in two flips is at least 91%. If Dominic makes $P(H) = K\%$, where K is an integer, what is the smallest value of K that meets his goal?

Problem 9

Consider the equation $mx^2 + 2x + m - 4 = 0$. For how many integers m does this equation have at least one negative root?

Problem 10

Laurie's favorite cookies are Oatmeal Raisin, Macadamia Nut, and Peanut Butter. Each day she packs 3 of these cookies for her lunch, sometimes bringing multiple cookies of the same type.

This long weekend her family is taking a 3-day trip, so Laurie needs to pack 3 days worth of cookies for her lunches. She packs 3 separate packs of cookies (one for each day) but packs them in no particular order. If no two packs of cookies are exactly the same, how many ways can Laurie pack and bring cookies for her trip?

Problem 11

A baseball diamond is a square with side length 90 feet. A catcher (labeled C) catches the ball at home base as an opposing player is running from 1st base to 2nd base. (See the diagram below for reference.)

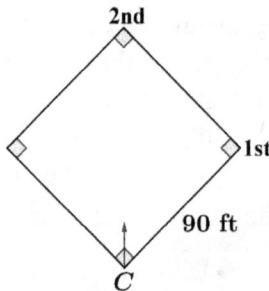

The catcher, who starts facing towards 2nd base, turns his head an angle halfway from 2nd to 1st base and sees the runner. How far, in feet, is the runner from 2nd base? Round your answer to the nearest multiple of 5 feet.

Problem 12
Felix has a box in the shape of a cube with volume 13824 cubic centimeters. His box has no top and he can rest a sphere on top of the box. If the sphere has a radius of 13 cm, how far is it from the bottom of the sphere to the bottom of the box (when the sphere is resting)? Round your answer to the nearest cm if necessary.

Problem 13
Calculate (using base 8) the following sum:

$$7_8 + 77_8 + 777_8 + 7777_8.$$

Here, for example, 77_8 denotes the two-digit base 8 number with digits 77. Input your answer in base 8 (without the subscript). For example, input 123_8 as 123.

Problem 14
Find all the rational solutions to the equation

$$\sqrt{2x^2 - 3x + 2} = \frac{2}{\sqrt{2x^2 - 3x + 2}} - 1.$$

What is the sum of these solutions, rounded to the nearest tenth?

Problem 15
How many integer pairs (x, y) are solutions to the equation

$$x^2 y - 2x^2 + 4y = 108?$$

Problem 16

An integer is randomly chosen from $1, 2, 3, \ldots, 21$ so that the probability of an integer k is proportional to k. For example, $P(2) : P(3) = 2 : 3$.

The probability that 21 is divisible by this number can be expressed as $\dfrac{A}{B}$ for positive integers A, B with $\gcd(A, B) = 1$. What is $B - A$?

Problem 17

If real numbers x, y satisfy the equation $x^4 + y^2 = 4x^2 - 4xy - 16$, then $x \cdot y = K$ for an integer K. What is K?

Problem 18

In a circle with radius 8 and center O, OA and OB are perpendicular radii. Chord AC intersects BO at a point D. If $DO : DB = 3 : 1$, find CD. Round your answer to the nearest tenth.

Problem 19

Let $ABCD$ be a trapezoid with $\overline{AB} \| \overline{CD}$ and $AB : CD = 3 : 2$. Let points E, F be on $\overline{AB}$ such that $AE = EF = FB$ and points G, H be the intersections (respectively) of $\overline{CE}$ and $\overline{CF}$ with $\overline{BD}$. If the area of $ABCD = 150$, what is the area of $\triangle CGH$? Round your answer to the nearest integer.

Problem 20

The five students Allie, Bruce, Carlos, Danny, and Eric each attend the same lecture and are assigned to work together the second day. Because it is a large class, only some of them had a chance to meet each other during the first lecture. (Assume if two students met, both remember that they met correctly.)

They notice that everyone had met at least one other person in the group, but only one triple of students all met each other. How many different ways could the 5 students have met the first day of class?

1.2   ZIML November 2018 Junior Varsity

Below are the 20 Problems from the Junior Varsity ZIML Competition held in November 2018.
The answer key is available on p.217 in the Appendix.
Full solutions to these questions are available starting on p.94.

Problem 1
Recall, for example, the notation 31_8 denotes the base 8 (octal) number with digits 31. That is,

$$31_8 = 3 \cdot 8^1 + 1 \cdot 8^0 = 24 + 1 = 25$$

when written in decimal.

Cameron is dressing up as a time traveling robot from 5 years in the future for Halloween and wants to write octal 3123 on his robot as part of his costume. However, being a robot he wants 3123_8 to be written in binary! What are the digits of this number when converted to binary?

Problem 2
8 classmates, 3 boys and 5 girls, line up for a photograph. Two of the boys stand together, but not all three. How many different arrangements of the classmates are there?

Problem 3

Points A, B, C, and D are arranged (in that order) around a circle such that $\widehat{AB} = 60°$, $\widehat{AC} = 160°$, and $\widehat{DA} = 140°$. (Thus, for example, arc $\widehat{AC}$ above contains B, while arc $\widehat{DA}$ above does not contain B or C.) Let E be the intersection of chords AC and BD. What is the ratio $\dfrac{[ABE]}{[CDE]}$? Round your answer to the nearest tenth. Here $[ABE]$ denotes the area of $\triangle ABE$.

Problem 4

Consider the equation

$$(x^2 + 5x - 3)^3 + 2x^2 + 10x = (x^2 + 5x - 3)^2 + 6.$$

How many real solutions exist for this equation?

Problem 5

A box has 10 identical red balls and K identical green balls. You choose 2 balls randomly, without replacement, from the box. What is the smallest positive integer K so that the probability you get at least one green ball is $\geq 80\%$?

Problem 6

For how many integer pairs (x,y) is it true that

$$5x^2 + y^2 + 8 \leq 4xy + 6x?$$

Problem 7

In $\triangle ABC$, $AB = 3\sqrt{3}+3$ and $\angle B = 60°$. A circle of radius 3 is inscribed inside $\triangle ABC$. The length of side AC can be written uniquely as $AC = \sqrt{K}+\sqrt{L}$ for integers $K < L$. What is $K+L$?

Problem 8

How many factors of 204020 are larger than 2018?

Problem 9

Consider the equation

$$\sqrt{x^2+2} = |x+2|+2.$$

What is the sum of all the real solutions to this equation, rounded to the nearest hundredth? If there are no solutions, input 0 as your answer.

Problem 10

3 circles (with the same center) of radius 1, 2, and 3 are each divided into 5 equal sectors. Regions are colored gray or white as in the diagram below.

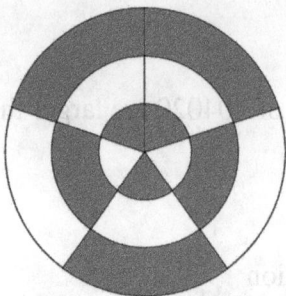

Problem 11

Let P be the collection of points that are the same distance from the point $(0,4)$ to the line $y = x$. (In fact, P forms a parabola.)

There is one point with integer coordinates (m,n) that is in P and also is on the line $y = -x$. What is $m \times n$?

Problem 12

What is the remainder when 11^{2018} is divided by 31?

Problem 13

How many ways are there to choose a, b, c, d, and e such that

$$a + b = d,$$
$$c + d = e,$$
$$\text{and } c + d + e = 18,$$

where a, b, c, d, and e are all non-negative integers?

Problem 14

Consider the set of points satisfying the equation

$$x^2 + (y - 10)^2 = \frac{1}{10}(y - 3x)^2.$$

Find the largest integer K such that any solution (x, y) in this set has $y \geq K$. What is K?

Problem 15

N is the smallest positive integer such that (i) N is a multiple of 1332, (ii) N is a multiple of 2376, and (iii) N has an odd number of factors. How many factors does N have?

Problem 16

An army operative parachutes from a plane when he is 2 miles directly above a radar station. Assume he drops in a straight line to a point 1 mile from the radar station on (flat) ground. $K\%$ of this path will be within a 1 mile distance from the radar station. What is K, rounded to the nearest integer?

Problem 17

The equation $2x^3 - 10x^2 + 13x - 3 = 0$ has three real roots a, b, and c, and $a > b > c$. If $a = 3$, what is the value of

$$4a^2 + b^2 + c^2 + 4ab + 4ac + 2bc?$$

Round your answer to the nearest integer.

Problem 18

Evaluate the sum

$$64\binom{6}{0} - 96\binom{6}{1} + 144\binom{6}{2} - 216\binom{6}{3}$$
$$+ 324\binom{6}{4} - 486\binom{6}{5} + 729\binom{6}{6}.$$

Problem 19

Two rectangles, $ABCD$ and $ABC'D'$ are formed in a square as shown in the diagram below.

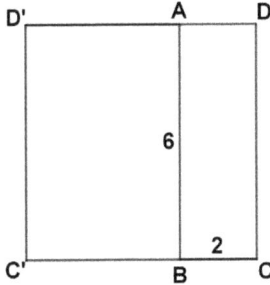

Triangles $\triangle ABE$ and $\triangle ABE'$ are formed such that AE, BE intersect CD at F, G and AE', BE' intersect $C'D'$ at F', G'. Assume $FG = 3$ and $F'G' = 2$. The ratio of the areas of $\triangle EFG$ to $\triangle E'F'G'$ can be expressed as $P : Q$ for positive integers P, Q with $\gcd(P, Q) = 1$. What is $P + Q$?

Problem 20

How many seven digit numbers of the form $\overline{20abc16}$ are divisible by 88?

1.3 ZIML December 2018 Junior Varsity

Below are the 20 Problems from the Junior Varsity ZIML Competition held in December 2018.
The answer key is available on p.218 in the Appendix.
Full solutions to these questions are available starting on p.106.

Problem 1
There are 3 sixth graders, 3 seventh graders, and 2 supervising teachers in a school club. The 8 line up for a photograph with none of the sixth graders next to each other. If the teachers stand outside the seventh graders, how many different photographs are possible?

Problem 2
$\triangle ABC$ has sides $AB = 20$, $BC = 21$, and $AC = 29$. $\triangle ABC$ is inscribed in a circle, and point D is constructed on the circle such that $\widehat{AD} = \widehat{DC}$ (with D not on arc $\widehat{ABC}$). The length BD can be written as $\dfrac{R\sqrt{S}}{T}$ for positive integers R, S, and T with S square-free and $\gcd(R,T) = 1$. What is $R + S + T$?

Problem 3
33, 66, and 99 are the first three examples of numbers divisible by 33 such that their digits (read left to right) are non-decreasing. What is the next smallest such number?

Problem 4
Consider the equation

$$\frac{x^4 - 5x^3 - x^2 + 17x + 12}{x^2 - 3x - 4} = 0.$$

What is the smallest real solution? Round your answer to the nearest tenth.

Problem 5
Recall for example, $123_8 = 1 \times 8^2 + 2 \times 8 + 3 \times 1$ (written in base 8).

The difference between 3456_{10} and 2345_{10} is $3456 - 2345 = 1111$. What is the difference between 3456_8 and 2345_8? (Give your answer converted to base 10.)

Problem 6
Jerry's favorite pizza place has a special deal on the weekends for a 3-topping pizza for $12.99. The deal even allows repeated toppings on the same pizza, chosen from their 8 different toppings.

This weekend, Jerry plans to watch an NBA game on Saturday and an NFL game on Sunday, and wants to buy a different pizza for each day. How many ways can Jerry order pizza?

Problem 7

$\triangle ABC$ has an area of 72 with $AB = 15$ and $AC = 12$. Let point D be on BC with AD the angle bisector of $\angle BAC$. If E is the midpoint of AB, what is the area of $\triangle BDE$? Round your answer to the nearest tenth if necessary.

Problem 8

N is the smallest integer that is (i) divisible by 378, (ii) a perfect square, and (iii) has at least 4 different primes as factors. How many factors does N have?

Problem 9

The polynomials below are missing some of their terms.

$$P(x) = x^3 - x^2 \cdots - 8$$
$$Q(x) = x^5 + 4x^4 \cdots + 192$$

Both $P(x)$ and $Q(x)$ have only real, integer roots and all the roots of $P(x)$ are roots of $Q(x)$. What is the smallest root of $Q(x)$?

Problem 10

Points A and B have coordinates $(4,2)$ and $(1,-7)$ respectively. Consider paths starting at point A, going to the line $y = x$, and then to point B. The length of the shortest such path can be written as $\sqrt{D}$ for an integer D.

Problem 11

A fair six-sided die numbered $1 - 6$ is rolled 4 times. Find the probability that two of the rolls are prime numbers, one is a composite number, and one roll is a 1 (which is neither prime nor composite). This probability can be written as $\dfrac{P}{Q}$ for positive integers P and Q with $\gcd(P, Q) = 1$. What is $Q - P$?

Problem 12

John was tasked with coming up with random two digit numbers. He starts with an initial number S. Given a current number C, he generates the next number N as the last two digits of $11 \times C + 5$.

Unfortunately John's numbers are not really random and all start repeating pretty quickly. In fact, for many S, the numbers alternate between only 2 numbers! For example, if $S = 2$, John's random numbers are

$$2, 27, 2, 27, \ldots.$$

How many initial numbers S from 0 to 99 have a pattern that alternates between 2 numbers? (Include the example $S = 2$ in your answer.)

Problem 13

A regular octagon shares a side with a square. The octagon's area is M times the area of the square. What is $\lfloor M \rfloor$? (Recall that $\lfloor x \rfloor$ is the greatest integer $\leq x$.)

Problem 14

The Fibonacci sequence starts $F_1 = F_2 = 1$ with $F_n = F_{n-1} + F_{n-2}$ for $n \geq 3$.

Kim creates a new sequence G_n where $G_n = 2 \cdot F_n - 2$. Kim notices that in fact G_n can also be given by a recursive definition:

$$G_1 = G_2 = 0, \text{ with } G_n = a \cdot G_{n-1} + b \cdot G_{n-2} + c \text{ for } n \geq 3.$$

(here a, b, and c denote real numbers). What is $a + b + c$? Round your answer to the nearest tenth.

Problem 15

Find the remainder when 2019^{2019} is divided by 45?

Problem 16
A point is chosen randomly in the square shown below.

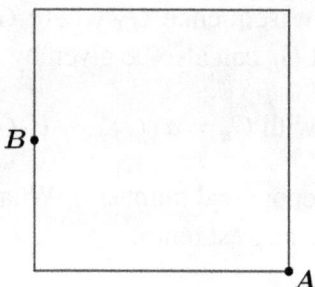

Point A is the lower right corner of the square and B is the midpoint of the left side. The probability that the chosen point is closer to A than to B can be written as $K\%$. What is K, rounded to the nearest tenth?

Problem 17
Consider the function

$$f(x) = \sqrt{6x - x^2} - \sqrt{4x^2 - 24x + 52}.$$

B is the smallest integer such that $f(x) \leq B$ for all x in the domain of $f(x)$. What is B?

Problem 18
Solve the equation $|x^2 + 6x + 9| + 6 = |5x + 15|$. What is the sum of all the real roots? Round your answer to the nearest tenth.

Problem 19

A right square pyramid has a base area of 36 and a total surface area of 96. Using a plane parallel to the base, the pyramid is cut, dividing the pyramid into two pieces with half the height of the original pyramid. The larger piece is shown below.

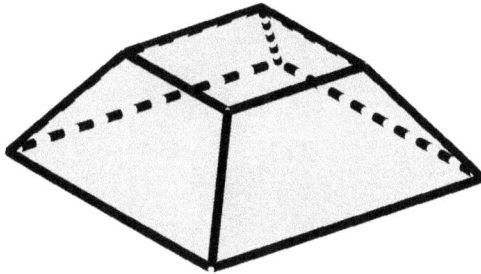

What is the surface area of this piece? Round your answer to the nearest integer.

Problem 20

For how many integers L do the graphs of

$$y = x^2 - 3x - 4 \text{ and } y = x^2 + Lx$$

intersect at (at least) one non-negative y value?

Problem 9

A right square pyramid has a base area of 30 and a total surface area of 90. Using a plane parallel to the base, the pyramid is cut, dividing the pyramid into two pieces with half the height of the original pyramid. The larger piece is shown below.

What is the surface area of this piece? Round your answer to the nearest integer.

Problem 10

For how many integers x do the graphs of

$$y = -x^2 - 3x - 4 \quad \text{and} \quad y = 4x + 1$$

intersect at (at least) one non-negative value?

1.4 ZIML January 2019 Junior Varsity

Below are the 20 Problems from the Junior Varsity ZIML Competition held in January 2019.
The answer key is available on p.219 in the Appendix.
Full solutions to these questions are available starting on p.118.

Problem 1
Among the perfect squares $1^2, 2^2, 3^2, \ldots, 99^2$, how many of them have odd numbers as the tens digits?

Problem 2
If $x^2 - 11x + 1 = 0$, find the value of $x^4 + x^{-4}$.

Problem 3
Define a recursive sequence with $a_4 = 10$ and for $n \geq 4$,

$$a_{n+1} = a_n + \binom{n}{n-2}.$$

What is a_{20}?

Problem 4
Let a and b be positive integers, where $a - b = 120$, and $\dfrac{\text{lcm}(a,b)}{\gcd(a,b)} = 105$. Find the value of $a + b$.

Problem 5

In acute triangle ABC, $\angle A > \angle B > \angle C$. Let α represent the smallest angle among $\angle A - \angle B$, $\angle B - \angle C$, and $90° - \angle A$. What is the maximum possible value for α (in degrees)?

Problem 6

Consider a pentagon and all its diagonals as shown below.

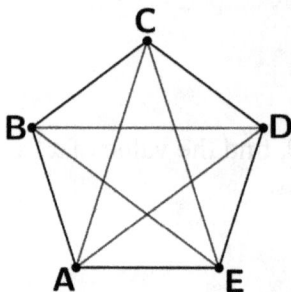

Consider paths that travel along the edges and diagonals of this pentagon, never traveling on the same line segment more than once. How many of these paths start and end at A, but do not visit any other vertex more than once?

Problem 7

Jackie is just starting out with the long jump, so she is very inconsistent. Each time she jumps, she randomly jumps somewhere between 0 and 2 meters (all distances are equally distributed). At her next track meet, Jackie will jump 3 times and her goal is to jump at least 4 meters in total. The probability Jackie meets her goal is $K\%$. What is K, rounded to the nearest integer.

Problem 8
Let $N = \overline{abcd}$ be a four-digit number, where a, b, c, and d are digits. Also assume that

$$N = (\overline{ab} + \overline{cd})^2,$$

in other words, N is the square of the sum of its own first two digits and last two digits. Find the sum of all such four-digit numbers N.

Problem 9
Let a and b be positive integers such that

$$1 \times 2 \times 3 \times \cdots \times 99 \times 100 = 12^a \times b,$$

find the maximum possible value of a.

Problem 10
The side length of square $ABCD$ is $\sqrt{3}$, let E be a point on the extension of $\overline{BC}$, F be the intersection of $\overline{AE}$ and $\overline{CD}$. Suppose $AF = CF + CE$, find length of $\overline{AF}$.

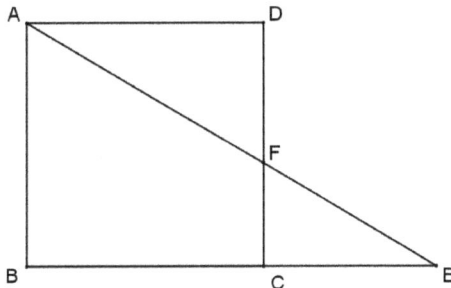

Problem 11

Let k be a real number, such that the equation

$$(x^2 - 1)(x^2 - 4) = k$$

has four nonzero real roots, and the roots form an arithmetic progression. Find the value of k, rounded to the nearest hundredth if necessary.

Problem 12

As shown in the diagram, $ABCD$ is a parallelogram, the circle through the three points A, B, and C is tangent to line $\overline{CD}$, and intersects $\overline{AD}$ at E.

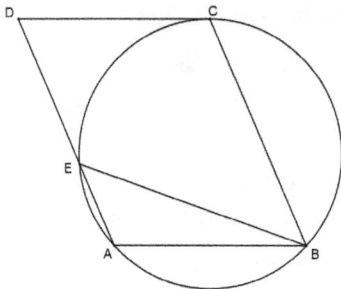

Given that $AB = 4$, and $BE = 5$, find the length of ED, rounded to the nearest tenth if necessary.

Problem 13

Phil just got 10 new golf balls as a holiday gift. He has 4 boxes in his garage where he keeps his golf balls. Phil decides to put at least one (and possibly all) of the new golf balls away into the boxes. (Phil does not necessarily use all the boxes.) If all the balls are identical, how many ways can Phil decide to put the golf balls into the boxes?

Problem 14

Start with regular hexagon H_0 with area 1. Recursively create regular hexagon H_{n+1} using the midpoints of the sides of hexagon H_n. What is the first n for which H_n has area ≤ 0.25?

Problem 15

The smallest root of

$$\frac{x+2}{x+1} - \frac{x+1}{2x+4} = \frac{17}{6}$$

can be written as $\dfrac{P}{Q}$ for integers P and $Q > 0$ with $\gcd(P,Q) = 1$. What is $P+Q$?

Problem 16

Raul was dealt 7 cards, one at a time. Without thinking he ordered them and noticed he had 5 numbers in a row, as well as 2 jacks. The cards he got were labeled

$$2, 3, 4, 5, 6, J, \text{ and } J.$$

Given that he got the 7 cards above, the probability that Raul was dealt the 2 before the 3, before the 4, etc., up to the 6 (we do not care when he was dealt the jacks) can be expressed as $\dfrac{P}{Q}$ for positive integers P and Q with $\gcd(P, Q) = 1$. What is $Q - P$?

Problem 17

Let a be a real number, and suppose the parabola

$$y = x^2 - 2ax + a^2 + 2a$$

and the line $y = 2x + 4$ intersect at two points A and B. Find the maximum possible length of $\overline{AB}$, rounded to the nearest integer if necessary.

Problem 18

Let x and y be positive real numbers such that

$$\frac{1}{x} - \frac{1}{y} = \frac{1}{x+y},$$

and suppose

$$\left(\frac{x}{y}\right)^3 + \left(\frac{y}{x}\right)^3 = P\sqrt{Q},$$

where P and Q are positive integers and Q is not a multiple of the square of any prime number, find the value of $P+Q$.

Problem 19

The general term of the sequence $\{a_n\}$ is defined as

$$a_n = \frac{1}{\sqrt{n}+\sqrt{n+1}}.$$

Let S_n represent the sum of the first n terms of the sequence, that is,

$$S_n = a_1 + a_2 + \cdots a_n.$$

Suppose $S_M = 18$. What is M?

Problem 20

Let x, y, and z be positive integers satisfying

$$\begin{cases} xy+yz = 45, \\ xz+yz = 17. \end{cases}$$

Find the maximum possible value of $x^2 + y^2 + z^2$.

1.5   ZIML February 2019 Junior Varsity

Below are the 20 Problems from the Junior Varsity ZIML Competition held in February 2019.
The answer key is available on p.220 in the Appendix.
Full solutions to these questions are available starting on p.133.

Problem 1
A group of people were surveyed and asked whether they regularly drink coffee and whether they regularly drink soda.

50% of them said they regularly drink coffee. 40% of them said they do not regularly drink soda. 70% of them said they regularly drink coffee or soda (or both).

$K\%$ of the people surveyed said they regularly drink coffee or soda but not both. What is K?

Problem 2
What is the smallest 7-digit number of the form $\overline{2020ABC}$ that is divisible by 75, 130, and 140?

Problem 3
From his basketball team, Luke has 4 different trophies and 7 identical basketballs. He wants to take a photo of his trophies lined up with the basketballs. If all the trophies are separated by at least one basketball, how many ways can Luke line up the objects?

Problem 4
The cubic equation $x^3 - 4x^2 + Cx - 6 = 0$ has 3 integer roots. What is C? Round your answer to the nearest tenth.

Problem 5
Polygon $ABCDEFGH$ is a regular octagon with side length 1. The area of quadrilateral $ACDH$, denoted as $[ACDH]$, can be expressed as $A + B\sqrt{C}$ for rational numbers A, B and a square-free integer C. What is $A + B + C$, rounded to the nearest hundredth.

Problem 6
How many factors of 180^{20} are not divisible by 10?

Problem 7
What is the minimum value of the function $f(x) = 4^x - 3 \cdot 2^x + 4$? Round your answer to the nearest hundredth.

Problem 8
The perimeter of a triangle is 30. If the triangle has distinct, integer side lengths, the maximum area of this triangle is K. What is $\lfloor K \rfloor$?

Recall $\lfloor x \rfloor$ is the greatest integer $\leq x$.

Problem 9
Let r and s be the two roots of $4x^2 + 16x + 9 = 0$. Then $r^3 + s^3$ can be written as $\dfrac{P}{Q}$ for integers P and Q with $Q > 0$ and $\gcd(P,Q) = 1$. What is $P + Q$?

Problem 10
Consider the number

$$N = 20192019\cdots 2019$$

where 2019 is written a total of nine times.

What is the remainder of N^{2019} when divided by 11?

Problem 11
Point A has coordinates $(-4, 1)$ and point B has integer coordinates (r,s). C has coordinates $(-2,4)$ and is on the perpendicular bisector of $\overline{AB}$. If $r > 0$, what is the smallest possible value of s?

Problem 12
Randomly pick real numbers a and b with $0 \le a,b \le 1$, determining the points $(0,a)$ and $(b,0)$.

The probability that the distance between the two points is at most $\dfrac{1}{2}$ is $\dfrac{R\pi}{S}$, for positive integers R and S with $\gcd(R,S) = 1$. What is $R \cdot S$?

Problem 13

In $\triangle ABC$, $AB = 5\sqrt{10}$, $AC = \sqrt{34}$, and $BC = 2\sqrt{26}$. Find the area of $\triangle ABC$, rounded to the nearest integer if necessary.

Problem 14

For a real number R, George expanded and simplified

$$(x^2 + R)^4 - (x+4)^6$$

to get a polynomial $P(x)$ with no x^4 term. What is the constant term of this polynomial?

Problem 15

Michael and Christopher were supposed to add up the terms of the geometric sequence

$$1, 3, 9, \ldots, 3^{20}$$

and then find the remainder when divided by 40.

Michael did the question correctly and got the answer M, but Christopher took the remainders of each term separately and then added all these remainders up to get the incorrect answer C. What is $C - M$?

Problem 16
Two different ways to write $35,937,000$ as the product of 4 positive integers include

$$8 \cdot 27 \cdot 125 \cdot 1331$$
$$\text{and } 1331 \cdot 8 \cdot 27 \cdot 125.$$

How many ways (including the two above) are there in total?

Problem 17
Rose has a fair coin and a fair 6-sided dice. She flips the coin, if she gets heads she throws the dice once, and if the gets tails she throws the dice three times.

The probability that the sum of the numbers in the dice is more than 4 is $\dfrac{P}{Q}$ for positive integers P and Q with $\gcd(P,Q) = 1$. What is $Q - P$?

Problem 18

In the circle below, $\overline{AB}$ is a diameter, $\overline{BC} \parallel \overline{DE}$, $\triangle CDG$ is isosceles, and $\widehat{BC}$ measures $60°$.

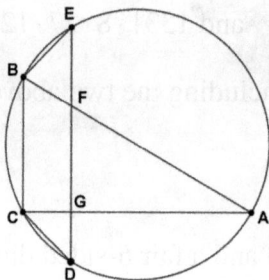

What is the measure of (minor) $\widehat{AE}$? Round your answer to the nearest integer.

(Recall a minor arc measures less than $180°$.)

Problem 19

Define a recursive sequence, where

$$a_{n+1} = \begin{cases} a_n \div 2, & \text{if } a_n \text{ is even,} \\ 3 \cdot a_n + 1, & \text{otherwise.} \end{cases}$$

If $a_1 = 21504$, what is the first M such that $a_M = 1$?

Problem 20

Consider integers K so that the equation

$$|Kx + 3| = x^2 + Kx + 1$$

has more than 2 real solutions. What is the product of all such values of K? Round your answer to the nearest tenth.

1.6 ZIML March 2019 Junior Varsity

Below are the 20 Problems from the Junior Varsity ZIML Competition held in March 2019.
The answer key is available on p.221 in the Appendix.
Full solutions to these questions are available starting on p.146.

Problem 1
How many factors does 234432 have?

Problem 2
Let $a \neq b$ denote the roots of $x^2 + 6x + 2 = 0$. What is $\dfrac{a^4 - b^4}{a - b}$?
Round your answer to the nearest integer.

Problem 3
Ramona has a deck of 9 cards numbered 1 through 9. She randomly removes one card from the deck and then picks one card from the ones remaining.

The probability that she picks a card with an odd number is $\dfrac{P}{Q}$ as a fraction in lowest terms. What is $Q - P$?

Problem 4
How many three-digit numbers $\overline{abc}$ are such that $\overline{ab} + c$ is a multiple of 11?

Problem 5

Let ABC be a triangle with $AB = 18$, $BC = 24$ and $CA = 30$. Let D be a point on BC such that $\angle CAD = \angle DAB$.

What is BD?

Problem 6

A point (a,b) is chosen uniformly at random where x and y are real numbers with $0 \le a, b \le 3$.

Harry picked the line $y = x + 1$ while Irene picked the line $y = x - 2$. The probability that (a,b) is closer to Irene's line than Harry's can be written as $\dfrac{P}{Q}$ for positive integers P and Q with $\gcd(P, Q) = 1$. What is $P + Q$?

Problem 7

The function $f(x) = \sqrt{4x - \sqrt{9 - x^2}}$ has domain all x such that $L \le x \le 3$. L can be written as $\dfrac{R\sqrt{S}}{T}$ for integers R, S, and $T > 0$ with $\gcd(R, T) = 1$ and S containing no squares as factors. What is $R + S + T$?

Problem 8

Right triangle $\triangle ABC$ with hypotenuse AB is inscribed in a circle with radius 10. Chords $\overline{BD}$ and $\overline{CE}$ bisect $\overline{AB}$ and $\overline{AC}$ respectively. Let F be the intersection of chords $\overline{BD}$ and $\overline{CE}$. If $BF = 8$ then DF can be written as $\dfrac{P}{Q}$ for positive integers P and Q with $\gcd(P, Q) = 1$. What is $P - Q$?

Problem 9
Consider the rational roots of

$$3x(3x+2) = 2|3x+1| - 1.$$

The sum of the squares of these rational roots can be written as $\dfrac{P}{Q}$ for positive integers P and Q with $\gcd(P,Q) = 1$. What is $P+Q$?

Problem 10
Ben rolls a 6-sided die once, noting the number $(1,2,\ldots,6)$. He then rolls the die that many times, counting the number of 1s, 2s, etc, which he records in a table. (He does not count the first roll in these results.) A few sample rows in this table are shown below.

1s	2s	3s	4s	5s	6s
2	0	0	0	0	0
0	1	3	1	1	0

How many different rows are possible in this table?

Problem 11
A recursive sequence satisfies the recurrence $x_{n+1} = 9 \cdot x_n + 1$. The last digit of x_{2019} is 4. If $x_0 = I$ is a one-digit integer, what is I?

Problem 12
What is the sum of all the real roots (multiplicities not included) of $x^4 - 3x^3 + 4x^2 - 3x + 1 = 0$?

Problem 13
Written in base 2, 3, or 5, the number N has a last digit (ones digit) of 0. Written in base 8, N has last two digits 22. If N is a three-digit number, what is N?

Problem 14
Points A, B, C, and D are collinear (in that order) with $AB = 4.4$, $BC = 5.6$, and $CD = 7.92$ (hence $AD = 17.92$). Point E is such that $AE = 8$ and $DE = 12.48$. If $\angle AEB = \angle DEC$ and $CE = 6$, what is the perimeter of $\triangle BCE$? Round your answer to the nearest hundredth.

Problem 15
The constant term in the expansion of

$$\left(x^3 + \frac{1}{x\sqrt{x}}\right)^n$$

is 84. What is n?

Problem 16
$p(x)$ is a degree 3 polynomial with leading coefficient 1. $p(x) \div (x - 1)$ has remainder 4, $p(x) \div x$ has remainder 6, and $p(x) \div (x + 1)$ has remainder 0. What is the largest real root of $p(x) = 0$? Round your answer to the nearest tenth.

Problem 17

The regular hexagonal pyramid shown below has a base with area $18\sqrt{3}$. The surface area of the entire pyramid is $48\sqrt{3}$.

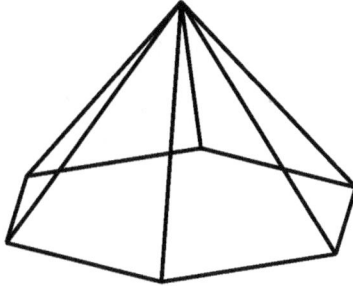

Problem 18

What is the remainder when 2017^{2018} is divided by 2019?

Problem 19

Define "M" numbers as follows:

- The "M" numbers consists of the digits 1, 2, and 3 only;
- The first digit is 1;
- The digits go up and down alternately.

For example, there are three 3-digit "M" numbers: 121, 131, and 132.

How many 12-digit "M" numbers are there?

Problem 20

In the following diagram the two smaller squares have area 9 and 16, and all three sides of the triangle have integer length.

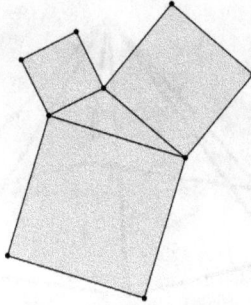

If the triangle has area $\dfrac{\sqrt{455}}{4}$, what is the area of the big square?

1.7 ZIML April 2019 Junior Varsity

Below are the 20 Problems from the Junior Varsity ZIML Competition held in April 2019.
The answer key is available on p.222 in the Appendix.
Full solutions to these questions are available starting on p.159.

Problem 1
16 people are seated evenly around a circular table. To choose teams for a game, 8 (identical) white cards and 8 (identical) black cards are distributed among the people. If each person gets the same color as the person seated directly across from them, how many ways can the cards be distributed?

Problem 2
A parabola with minimum at its vertex $(-1, -3)$ intersects a parabola with maximum at its vertex $(2, 1)$ at exactly one point. If the leading coefficients of both parabolas have the same absolute value, then this absolute value can be written as $\dfrac{P}{Q}$ for positive integers P and Q with $\gcd(P, Q) = 1$. What is $P + Q$?

Problem 3

After learning about factors Terry practiced counting factors. To document his process he made a table of all the integers from 1 to 100 and how many factors each had:

Number	Number of Factors
1	1
2	2
$\vdots$	$\vdots$

If Terry adds up all the odd numbers in the Number of Factors column, what is the result?

Problem 4

Consider positive integers B such that $x = \dfrac{1}{2} \pm \dfrac{\sqrt{B}}{2}$ are solutions to $(x+4)(x-5) + 5 = \dfrac{14}{(x+1)(x-2)}$. What is the sum of all possible values of B? Input an answer of 0 if there are no possible values of B.

Problem 5

$\overline{AB}$ and $\overline{CD}$ are diameters of a circle of radius 6. Arcs $\overset{\frown}{AC}$ and $\overset{\frown}{BD}$ both measure $60°$. The area of the region bounded by $\overset{\frown}{AC}$, $\overset{\frown}{BD}$, AD, and BC can be written as $R\pi + S\sqrt{T}$ for integers R, S, and T with T containing no squares as factors. What is $R + S + T$?

Problem 6

Let $x = \dfrac{1}{2}(\sqrt{13} - 3)$. Calculate

$$x(x+1)(x+2)(x+3).$$

Round your answer to the nearest integer.

Problem 7

A and B are randomly chosen from the factors of 2000. (A and B are not necessarily distinct.) The probability that $\gcd(A,B) = 1$ can be written as $\dfrac{P}{Q}$ for positive integers P and Q with $\gcd(P,Q) = 1$. What is $Q - P$?

Problem 8

Quadrilateral $ABCD$ is inscribed in a circle, with $AB = 5\sqrt{2}$, $AC = 3\sqrt{10}$, and $AD = 5\sqrt{2}$. If $\triangle BAD$ has area 25, what is the area of $\triangle BCD$? Round your answer to the nearest integer.

Problem 9

Recall a lattice point is a point (x,y) with integer coordinates.

How many lattice points (x,y) lie on the curve $y = \dfrac{x^2 + 14}{15}$ with $0 \le x, y \le 100$?

Problem 10

The graphs of $y = \big||x| - 6\big|$ and $y = mx + 4$ intersect exactly 4 times for all real numbers m in the range $L < m < U$. What is $U - L$? Round your answer to the nearest tenth.

Problem 11

Patrick went to the local bakery to buy some cupcakes. The bakery sells chocolate, vanilla, and fudge cupcakes. Each type is available with or without frosting. How many collections of 8 cupcakes can Patrick buy, if he buys at least one cupcake with frosting?

Problem 12

Recall a palindrome is a number that is the same read left-to-right or right-to-left.

There are no 5-digit palindromes divisible by the first 6 primes, but there is one 5-digit palindrome divisible by the primes if we exclude 5. What is this 5-digit palindrome?

Problem 13

In $\triangle ABC$, medians $\overline{AD}$ and $\overline{BE}$ are extended to points J and K, respectively, such that $AD : AJ = BE : BK = 3 : 4$ (with D on $\overline{AJ}$ and E on $\overline{BK}$). The ratio of areas $[ABC] : [CJK]$ can be written as $P : Q$ for positive integers P and Q with $\gcd(P, Q) = 1$. What is $Q - P$?

Problem 14

What is the remainder when

$$1 + 8 + 27 + \cdots + 2019^3$$

is divided by 2021?

Problem 15

A circle of radius $\sqrt{3}$ is tangent to a given line at N. Line segment $\overline{AB}$ has A on this line, tangent to the circle at its midpoint M. Line segment $\overline{CD}$ has C on this line, with B its midpoint, as in the configuration shown below.

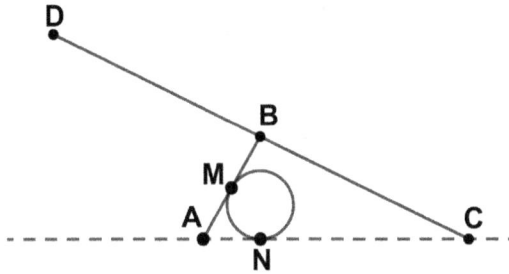

If $AB = 6$ and $CD = 24$, then CN can be written as $\sqrt{R}$ for a positive integer R. What is R?

Problem 16

Sara has 15 cards numbered $1, \ldots, 15$. She chooses two cards at random without replacement. The probability that the product of the two numbers on her cards is a multiple of 6 is $\dfrac{P}{Q}$ as a simplified fraction. What is $Q - P$?

Problem 17

Square $ABCD$ has coordinates $A = (0,0)$, $B = (1,0)$, $C = (1,1)$, and $D = (0,1)$. $\triangle AEF$ is formed, as shown below, with E the midpoint of $\overline{BC}$ and F the midpoint of $\overline{CD}$.

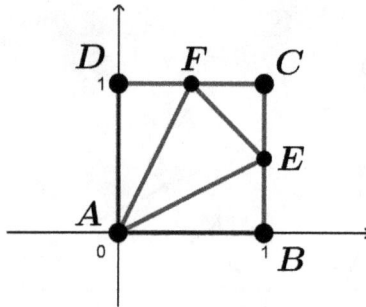

Square $A'B'C'D'$ is the image of $ABCD$ under a dilation with scale factor 2 centered at $(0,0)$. (Thus $A' = (0,0)$, $B' = (2,0)$, $C' = (2,2)$, $D' = (0,2)$.) If $A''E''F''$ is the image of AEF under a dilation with scale factor 3 centered at $(0,0)$, then the fraction of $A''E''F''$ inside $A'B'C'D'$ can be expressed as $\dfrac{P}{Q}$ for positive integers P and Q with $\gcd(P,Q) = 1$. What is $Q - P$?

Problem 18
Find the last two digits of 23^{2323}.

Problem 19
Linus randomly picks a number M from $1,2,3,4,5$ and draws the line $y = Mx - 1$. Polly picks a point (x, y) with $1 \leq x \leq 3$ and $0 \leq y \leq 2$. The probability that Polly's point is above Linus's line can be written as $K\%$. What is K, rounded to the nearest hundredth?

Problem 20
The cubic polynomial $2x^3 - x^2 - 13x - 6 = 0$ has three real roots, call them r, s, and t. The ratio $\dfrac{r^2 + s^2 + t^2}{r + s + t}$ can be written as $\dfrac{P}{Q}$ for integers P and Q with $Q > 0$ and $\gcd(P, Q) = 1$. What is $P + Q$?

1.8   ZIML May 2019 Junior Varsity

Below are the 20 Problems from the Junior Varsity ZIML Competition held in May 2019.
The answer key is available on p.223 in the Appendix.
Full solutions to these questions are available starting on p.173.

Problem 1
What is the remainder when 2^{503} is divided by 30?

Problem 2
Consider the degree 10 polynomial

$$p(x) = (x^2 + 6x + m)(x^2 + 7x + m)(x^2 + 8x + m) \cdot (x^2 + 9x + m)(x^2 + 10x + m).$$

For how many positive integers m does $p(x)$ have 10 distinct real zeros?

Problem 3

Triangle $\triangle ABC$ has sides of length $AB = 15$, $BC = 26$, and $AC = 37$. The angle bisector $\overline{AD}$ is drawn, as well as squares with side length BD and CD (outside the triangle). This creates the heptagon shown below.

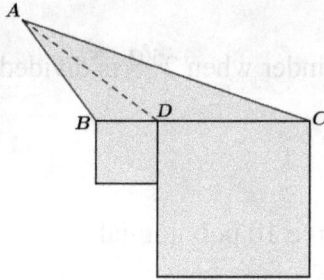

What is the area of this heptagon? Round your answer to the nearest hundredth.

Problem 4

Pete has to pick a new pin for his online shopping account. The pin consists of 5-digits, each chosen from 0 to 9. Pete chooses a pin so that the 5 digits are all increasing, but not all consecutive. (Thus, 01234 is not allowed, but 01245 is.) How many different pins could Pete have chosen?

Problem 5

Two circles, with radii of 2 and 1 respectively, share the same center O. Diameters $\overline{AB}$ and $\overline{CD}$ of the two circles are collinear with $AB > CD$. E lies on the line tangent to the larger circle at B with $EO = 4$. Then the area of $\triangle CDE$ can be expressed as $R\sqrt{S}$ for positive integers R and S, where S contains no squares as factors. What is $R + S$?

Problem 6

Consider the function $f(x) = x - 4\sqrt{x-4} - 13$. The minimum value of this function is M, occurring when $x = L$. What is $L + M$? Round your answer to the nearest tenth.

Problem 7

Consider the set of all the 10-digit integers that are multiples of 9. For each number in this set, calculate the sum of its digits. How many different sums are possible?

(Recall $1,000,000,000$ is the smallest 10-digit integer.)

Problem 8

In convex pentagon $ABCDE$, $AB = BC = CD$ with $\angle A = \angle B = \angle C = \angle D = 120°$. If quadrilateral $ABCD$ has area $27\sqrt{3}$, then the area of the full pentagon can be written as $K\sqrt{3}$ for an integer K. What is K?

Problem 9

The quadratic equation $x^2 + 4x + C = 0$ has roots $p < q$. The quadratic equation $x^2 - 20x + D = 0$ has roots $r < s$. If p, q, r, s forms an arithmetic sequence, what is D?

Problem 10

Sam looks at the triangle bounded by the lines $y = \frac{1}{2}x$, $y = -x$, and $y = 2x - 4$. He calculates the ratio of its area to is perimeter, getting an answer of s as a fraction. Jon does the same with the triangle bounded by the lines $y = \frac{1}{2}x$, $y = -x$, and $y = 2x - 15$, getting an answer of t as a fraction.

$\frac{s}{t}$ can be written as $\frac{P}{Q}$ for positive integers P and Q with $\gcd(P, Q) = 1$. What is $P + Q$?

Problem 11

A real number R is chosen randomly with $0 \leq R \leq 150$. The probability that R is closer to an odd perfect square (for example, 1 or 81) than an even perfect square (for example, 0 or 100) is $K\%$. What is K, rounded to the nearest integer?

Problem 12

$\triangle ABC$ is a right triangle with integer side lengths and an area less than 100. If $\triangle ABC$ has the largest possible area, what is its area?

Problem 13
From the set $\{0,1,2,3,4,5,6,7,8,9\}$, select 3 of them so that the sum of the 3 selected numbers is an even number and is at least 10. How many ways are there to choose these 3 numbers?

Problem 14
x is a real number satisfying the cubic equation $x^3 + 5x - 3 = 0$. Calculate $x^5 + 8x^3 - 3x^2 + 15x$. Round your answer to the nearest integer.

Problem 15
How many integers N from 1 to 1000 satisfy $\gcd(N, 100) > 1$?

Problem 16
Pauli has 5 identical black cards and 7 different red cards. He starts by lining up the 5 black cards, and then places 4 of the red cards in-between the black cards (one in-between each pair). Lastly, one by one, he places the remaining red cards next to one of the red cards already placed. How many different arrangements can Pauli achieve using this method?

Problem 17
The equation $8^x - 16 \cdot 4^x + 60 \cdot 2^x - 64 = 0$ has three real solutions. What is the sum of these solutions?

Problem 18

A perfect square N is made from the digits 0, 2, 3, 4, and 9. If N is also divisible by 22, what is N?

Problem 19

A sphere has a radius of 1. Points A, B, and C are on the sphere with arc lengths $\widehat{AB} = \widehat{AC} = \dfrac{\pi}{2}$ and $\widehat{BC} = \dfrac{\pi}{4}$. Further, both arcs $\widehat{AB}$ and $\widehat{AC}$ meet $\widehat{BC}$ at a right angle. The surface area of the sphere contained between the three arcs $\widehat{AB}$, $\widehat{AC}$, and $\widehat{BC}$ can be expressed as $M \cdot \pi$ for a real number M. What is m rounded to the nearest hundredth?

Problem 20

John and his 5 friends have a movie night and decide to watch two movies. They decide who gets to pick the first movie according to the following odds:

John	Iris	Helen	Greg	Fran	Erin
25%	25%	25%	10%	10%	5%

They also randomly choose, using the same odds again, who gets the second pick. However, the person chosen for the first pick cannot also get the second pick. (In the event that the same person is chosen, they pick again until a new person gets second pick.)

The probability that Erin gets to pick one of the movies can be expressed as $\dfrac{P}{Q}$ for positive integers P and Q with $\gcd(P,Q) = 1$. What is $Q - P$?

1.9 ZIML June 2019 Junior Varsity

Below are the 20 Problems from the Junior Varsity ZIML Competition held in June 2019.

The answer key is available on p.224 in the Appendix.

Full solutions to these questions are available starting on p.186.

Problem 1
Consider the set S of all x such that

$$f(x) = \sqrt{-x^2 - 3x + 10} - \sqrt{-x^2 + x + 6}$$

is defined and $f(x) \geq 0$. Set S can be written as the closed interval $[L, U]$, for integers L and U. What is $L \cdot U$?

Problem 2
A school's JV Cross Country team has 10 members in middle and high school. Among the middle school members there are 3 girls and 1 boy and among the high school members there are 4 boys and 2 girls.

Two middle school members and two high school members are chosen at random. The probability that there are 3 girls and 1 boy chosen can be written as $\dfrac{P}{Q}$ for positive integers P and Q with $\gcd(P, Q) = 1$. What is $P + Q$?

Problem 3
A circle is inscribed in a triangle with sides 16, 30, and 34. The area of this circle can be written as $K \cdot \pi$. What is K, rounded to the nearest integer?

Problem 4

A sequence has initial value $a_0 = 5$ and satisfies the recurrence

$$a_{n+1} = 4 \cdot a_n + 3.$$

What is the remainder when a_{1771} is divided by 17?

Problem 5

Peter and Paul are both planning on going to the gym tomorrow morning. Peter always works out for exactly 1 hour. Paul always leaves the gym at the start of an hour (9am, 10am, etc.) and works out at least 30 minutes but always less than 90 minutes.

Assume Peter and Paul arrive randomly between 8am and 11am (and their arrival times are independent from each other). The probability they are both at the gym at the gym the same time is $K\%$. What is K, rounded to the nearest integer?

Problem 6

A trapezoid $ABCD$ is drawn around two 5-12-13 right triangles, as shown in the diagram below.

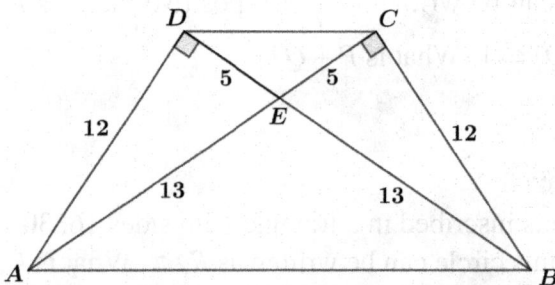

Problem 7

M is a two-digit integer such that $M > 10$ and $53_M = M_{53}$. That is, the (two-digit) base M number with digits 5 and 3 is the same as the (two-digit) base 53 number with the same digits as M. If M and 53 have no digits in common, what is M?

Problem 8

The graphs of $y = 2x + 3$ and $y = |x^2 - 2|$ intersect at 3 points. What is the product of the three x-coordinates where they intersect?

Problem 9

$ABCD$ is a quadrilateral with $AB = 2\sqrt{2}$, $BC = 2\sqrt{10}$, $CD = 4\sqrt{3}$, and $AD = 4$. Its diagonals $\overline{AC}$ and $\overline{BD}$ are perpendicular and intersect at point E. If $AE = 2$, then the area of $ABCD$ is S. What is S, rounded to the nearest integer?

Problem 10

Solve the equation

$$(2x^2 - 7x + 5)^2 - 2(2x^2 - 7x + 5)(10 - x) - 40x = 0.$$

What is the sum of all the real solutions? Round your answer to the nearest tenth.

Problem 11

Consider nine digit numbers with digits $\overline{6a6b6c6d6}$ that are divisible by 11. What is the sum of all the possible remainders when these numbers are divided by 9?

Problem 12

Sandra buys 10 identical birds and plans to place them in 4 (different) cages in her house. How many ways can she place them into the cages so that none of the cages have more than 4 birds?

Problem 13

Recall the difference of two squares is

$$a^2 - b^2 = (a+b)(a-b).$$

Brad and Claire both factored the expression $x^{192} - y^{192}$. Brad factored the expression as far as he could using only the difference of two squares. Claire first factored the expression as

$$(x^{64} - y^{64})(x^{128} + x^{64}y^{64} + y^{128})$$

and then as far as she could using only the difference of two squares.

Brad ended with an expression that was the product of P polynomials while Claire had an expression that was the product of Q polynomials. What is $P+Q$?

Problem 14

Consider the number N with prime factorization

$$N = 2^8 \cdot 3^7 \cdot 5^6 \cdot 7^5 \cdot 11^4 \cdot 13^3.$$

How many factors of N are perfect cubes that end in 3 zeros?

Problem 15

An arithmetic sequence and a geometric sequence both start with the first two terms 24 and 36. The terms of the arithmetic sequence (including 24 and 36) that are less than 100 are written on red cards while the terms of the geometric sequence (including 24 and 36) that are less than 100 are written on blue cards.

Let N be the number of ways to arrange all these red and blue cards in a line, so that no two blue cards are next to each other. The ratio $N : 10!$ is proportional to $P : Q$ for positive integers P and Q with $\gcd(P,Q) = 1$. What is $P+Q$?

Problem 16

A cube and a sphere share the same center. Each edge of the cube is half contained in the sphere. The ratio of the volume of the sphere to the cube can be written as $A\pi : B$ for positive integer A and B with $\gcd(A,B) = 1$. What is $A+B$?

Problem 17

There are infinitely many integer pairs (x,y) satisfying $x^3 - 4y = 9$. How many pairs are there such that $0 \leq x, y \leq 100$?

Problem 18

Among the integers $1, 2, \ldots, 1000$, how many are divisible by 6, 10, or 15, but not by all three at once?

Problem 19

The first four terms of the geometric sequence $1, r, r^2, r^3$ are roots of the polynomial

$$64x^4 + 696x^3 - 5510x^2 - 10875x + 15625 = 0.$$

r can be expressed as $\dfrac{P}{Q}$ for integers P and Q with $Q > 0$ and $\gcd(P, Q) = 1$. What is $Q - P$?

Problem 20

Right triangle $\triangle ABC$ with hypotenuse AC has points A, B, and C on the parabola $y = x^2$. If A is the vertex of the parabola and $AB = 2\sqrt{5}$, what is the area of $\triangle ABC$? Round your answer to the nearest hundredth.

2. ZIML Solutions

This part of the book contains the official solutions to the problems from the nine Junior Varsity ZIML Contests from the 2018-19 School Year.

Students are encouraged to discuss and share their own methods to the problems using the Discussion Forum on ziml.areteem.org.

2.1 ZIML October 2018 Junior Varsity

Below are the solutions from the Junior Varsity ZIML Competition held in October 2018.

The problems from the contest are available on p.17.

Problem 1 Solution

When expanded, the terms (before simplifying) containing x^5 are $(x^5)(1) = x^5$, $(-x^4)(2x) = -2x^5$, $(-x^3)(4x^2) = -4x^5$, and $(-x)(x^4) = -x^5$. Thus the coefficient is $1 - 2 - 4 - 1 = -6$.

Answer: -6

Problem 2 Solution

There are $4! = 24$ ways to arrange the four grades (freshman, sophomore, etc.). However, we still need to arrange the 3 students in each grade. This can be done in $3! = 6$ ways for each grade. Hence the total number of arrangements is $4! \cdot (3!)^4 = 24 \cdot 6^4 = 31104$.

Answer: 31104

Problem 3 Solution

The prime factorization of $308 = 2^2 \cdot 7 \cdot 11$, so we ensure the number is divisible by 4, 7, and 11.

To be divisible by 4, the last two digits $\overline{5B}$ must be divisible by 4. This implies either $B = 2$ or $B = 6$.

For divisibility by 11, the alternating sum of the digits $B - 5 + 2 - 5 + A - 1 = A + B - 9$ must be a multiple of 11. If $B = 2$ we have $A - 7$ is a multiple of 11, so $A = 7$ and similarly if $B = 6$ we have $A = 5$.

Hence our two possible numbers are 175252 and 155252. Manually checking we see only 175252 works, so is our number.

Answer: 175252

Problem 4 Solution

Note in fact adding up the 25 elements in the table is exactly $(r+s+t+u+v)^2$. By Vieta's theorem, $r+s+t+u+v = \dfrac{-11}{3}$.

Hence $(r+s+t+u+v)^2 = \left(-\dfrac{11}{3}\right)^2 = \dfrac{121}{9}$ so $N+M = 130$.

Answer: 130

Problem 5 Solution

If K is a multiple of $66 = 2 \cdot 3 \cdot 11$ it is divisible by 2, 3, and 11. For K to have an odd number of digits, it must be a perfect square, hence each exponent in its prime factorization must be even.

Combining these two restrictions gives the smallest number of $2^2 \cdot 3^2 \cdot 11^2 = 4356$. However, this number does not have repeated digits. We try multiplying by the smallest perfect square $(2^2 = 4)$ to get $4356 \cdot 4 = 17424$, which does have repeated digits. Therefore $K = 17424$.

Answer: 17424

Problem 6 Solution

In general, the sum of the interior angles of a convex polygon are $180°(n-2)$. The polygon described above has sum of interior angles

$$30° \cdot (1+2+\cdots+n) = 30° \cdot \frac{n(n+1)}{2}.$$

Therefore $180(n-2) = 15n(n+1)$ or $n^2 - 11n + 24 = 0$. Factoring we have $(n-3)(n-8) = 0$ so $n = 3$ or $n = 8$.

However, our polygon n must be convex. For $n = 8$ this polygon would have angles of $30°, 60°, \ldots, 240°$, so this is NOT a convex polygon. (Further, it also has a $180°$ angle, thus might be considered to have 7 sides.)

Therefore $n = 3$ is the only possibility, so the sum of all possible n is 3.

Answer: 3

Problem 7 Solution

By Fermat's Little Theorem, $x^{12} \equiv 1 \pmod{13}$, for any integer x not a multiple of 13. Thus for any integer $k \geq 0$, $x^{12k+1} \equiv x \pmod{13}$. Hence, none of Mr. David's number will be changed when $N + 5$ is $12k + 1$ for integers $k \geq 0$.

The smallest 3-digit multiple of 12 is 108, thus $N + 5 = 109$ so $N = 104$.

Caution: It is not immediately clear that $N = 104$ is the smallest. Fermat's Little Theorem guarantees that $x^{12} \equiv 1 \pmod{13}$, but it is possible for a smaller power to work as well. For example, $3^3 \equiv 1 \pmod{13}$. However, we can double check that 2^{12} is the first power of 2 to have remainder 1 when divided by 13, so $N = 104$ is the smallest.

Answer: 104

Problem 8 Solution

Let $P(H) = x$. If we want the probability of at least one heads in two flips to be $\geq 91\%$, then we need the probability of 0 heads to be $< 9\%$. Hence $(1 - x)^2 < 9\% = 0.09$. Therefore

$$(1-x)^2 < \frac{9}{100} \Rightarrow (1-x) < \frac{3}{10} \Rightarrow x \geq \frac{7}{10} = 70\%.$$

Therefore $K = 70$.

Answer: 70

Problem 9 Solution

To have at least one negative root, the equation must have real solutions, so the discriminant $\Delta > 0$. We have

$$\Delta = 2^2 - 4 \cdot m \cdot (m-4) = -4m^2 + 16m + 4 > 0 \Rightarrow m^2 - 4m - 1 < 0.$$

$m^2 - 16m - 4 = 0$ when $m = \dfrac{4 \pm \sqrt{16+4}}{2} = 2 \pm \sqrt{5}$. As $2 < \sqrt{5} < 3$, the equation $mx^2 + 2x + m - 4 = 0$ has real solutions when $m = 0, 1, 2, 3, 4$.

When $m = 0$, the equation is in fact linear, with root $x = 2$. Else $0 < m \leq 4$ and by Vieta's theorem the equation has sum of roots $-2/m < 0$ and product of roots $(m-4)/m > 0$ so at least one root is negative. Therefore, $m = 1, 2, 3, 4$ work, so our answer is 4.

Answer: 4

Problem 10 Solution

Laurie has 3 favorite types of cookies, so using stars and bars there are $\dbinom{3+3-1}{3} = \dbinom{5}{3} = 10$ ways for her to choose a collection of 3 cookies.

From these 10 collections, Laurie must choose 3 to bring on her trip. Therefore she has $\dbinom{10}{3} = 120$ total ways to pack and bring cookies.

Answer: 120

Problem 11 Solution

The runner's position to the catcher forms the angle bisector as seen below.

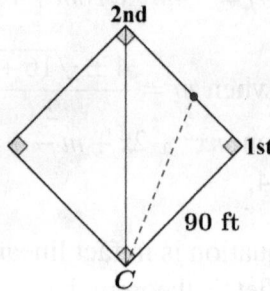

Labeling 1st base A, 2nd base B, the catcher C, and the runner's position R we have, using the angle bisector theorem,

$$\frac{RB}{RA} = \frac{CB}{CA} = \frac{90\sqrt{2}}{90} = \sqrt{2}$$

as CB is the diagonal of a square with side length 90. We know $RB + RA = 90$, so

$$RB + \frac{RB}{\sqrt{2}} = RB\left(\frac{2+\sqrt{2}}{2}\right) = 90$$

$$\Rightarrow RB = \frac{180}{2+\sqrt{2}} = 180 - 90\sqrt{2}.$$

Using the approximation $\sqrt{2} \approx 1.4$ we have $RB \approx 180 - 136 = 54$. Hence rounded to the nearest multiple of 5 we have the runner is approximately 55 feet from 2nd base. (The exact answer is ≈ 52.7208 feet.)

Answer: 55

Problem 12 Solution

The sphere is resting on the box, which has a side length of $\sqrt[3]{13824} = 24$ cm. Therefore the sphere is tangent to the four midpoints of the top edges of the box. Hence, we have the following side view of the sphere and box:

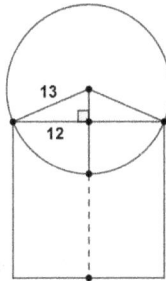

Using the Pythagorean theorem, the remaining side of the right triangle shown is 5 cm. As this gives the height of the center of the sphere above the sphere, the distance from the bottom of the sphere to the bottom of the box (the dotted line in the above diagram) is $24 + 5 - 13 = 16$ cm.

Answer: 16

Problem 13 Solution

It is not hard to do the calculation directly in base 8.

However, note $7_8 = 10_8 - 1_8$, $77_8 = 100_8 - 1_8$, etc. Thus

$$7_8 + 77_8 + 777_8 + 7777_8$$
$$= (10_8 - 1_8) + (100_8 - 1_8) + (1000_8 - 1_8) + (10000_8 - 1_8)$$
$$= 11110_8 - 4_8$$
$$= 11104_8$$

Thus our answer is 11104.

Answer: 11104

Problem 14 Solution

Letting $y = \sqrt{2x^2 - 3x + 2}$ we have $y = \dfrac{2}{y} - 1$ or $y^2 + y - 2 = 0$.
Therefore $(y+2)(y-1) = 0$ so $y = -2$ or $y = 1$.

$\sqrt{2x^2 - 3x + 2} \neq -2$, so we have $\sqrt{2x^2 - 3x + 2} = 1$ or $2x^2 - 3x + 1 = 0$. Factoring, $(2x - 1)(x - 1) = 0$ so $x = 0.5$ or $x = 1$. The sum of these two rational solutions is 1.5.

Answer: 1.5

Problem 15 Solution

Completing the rectangle we have

$$x^2 y - 2x^2 + 4y = (x^2 + 4)(y - 2) + 8.$$

Thus $(x^2 + 4)(y - 2) = 100$. If x, y are integers, so are $x^2 + 4$ and $y - 2$. Hence looking at factors of 100 we have

$$x^2 + 4 = 1, 2, 4, 5, 10, 20, 25, 50, 100$$
$$\Rightarrow \quad x^2 = -3, -2, 0, 1, 16, 21, 46, 96$$

so if x is an integer we have $x = 0$, $x = \pm 1$, or $x = \pm 4$. Each of these values of x gives a corresponding y value, so there are $1 + 2 + 2 = 5$ pairs of integer solutions.

Answer: 5

Problem 16 Solution

If we set $P(1) = x$, then $P(2), P(3), \ldots = 2x, 3x, \ldots$. As the probabilities must add to 1 we have

$$P(1) + P(2) + \cdots P(21) = x + 2x + \cdots + 21x = \frac{21 \cdot 22}{2} x = 1.$$

Therefore $P(1) = x = \dfrac{1}{231}$. As 21 is divisible by 1, 3, 7, and 21, the probability that 21 is divisible by the number chosen is

$$\frac{1}{231} + \frac{3}{231} + \frac{7}{231} + \frac{21}{231} = \frac{32}{231}.$$

This fraction is irreducible, hence $B - A = 231 - 32 = 199$.

Answer: 199

Problem 17 Solution

Rewriting we have $x^4 - 4x^2 + 16 + y^2 + 4xy = 0$. Adding and subtracting $4x^2$ we have

$$x^4 - 8x^2 + 16 + y^2 + 4xy + 4x^2 = (x^2 - 4)^2 + (y + 2x)^2 = 0.$$

(The motivation is now we have factored our expression as the sum of two squares.) Thus $x^2 - 4 = 0$ and $y + 2x = 0$. Hence $x = 2$ and $y = -4$ or $x = -2$ and $y = 4$. In both cases, $x \cdot = -8$ so $K = -8$.

Answer: -8

Problem 18 Solution

The circle has a radius 8 so since $DO : DB = 3 : 1$ we have $DO = 6$ and $DB = 2$. Using the Pythagorean theorem we know $AD = 10$.

Let BF be the diameter. Using power of a point we have

$$BD \cdot DF = AD \cdot DC \Rightarrow 2 \cdot 14 = 10 \cdot DC.$$

Therefore $CD = 28 \div 10 = 2.8$ as needed.

Answer: 2.8

Problem 19 Solution

As $\overline{AB} \| \overline{CD}$, $AB : CD = 3 : 2$, and $AE = EF = FB$, we have that $EBCD$ is a parallelogram. Therefore, G, which is the intersection of diagonals $\overline{BD}$ and $\overline{EC}$, is the midpoint of G.

Therefore, examining $\triangle BCE$, we have $\overline{CF}$ and $\overline{BG}$ are medians, hence H is the centroid of $\triangle BCE$. As the medians in a triangle divide it into 6 equal pieces, we have $[CGH] = \dfrac{1}{6} \cdot [BCE]$ (here we use $[CGH]$, for example, to denote area).

Since $ABCD$ is a trapezoid, the altitudes of $\triangle BCE$, $\triangle CED$, and $\triangle ADE$ are the same, so looking at their bases we have

$$[BCE] : [CED] : [ADE] = 2 : 2 : 1.$$

From this it follows that $[BCE] = \dfrac{2}{5} \cdot [ABCD]$.

Therefore

$$[CGH] = \frac{1}{6} \cdot [BCE] = \frac{1}{6} \cdot \frac{2}{5} \cdot [ABCD] = \frac{1}{15} \cdot [ABCD].$$

Thus, $[CGH] = 150 \div 15 = 10$.

Answer: 10

Problem 20 Solution

Label the students A, B, C, D, and E. There are $\dbinom{5}{3} = 10$ different triples of students that could all know each other. Without loss of generality, assume A, B, and C have all met each other. Consider two cases based on whether or not D and E met the first day.

First assume D and E did not meet. In this case, each of them must met at least one of A, B, C. In fact, they must have met exactly one of A, B, C, else a second triple of students would all have met. Hence there are 3 choices each for D and E. This gives $3 \cdot 3 = 9$ outcomes in this case.

If D and E did meet, they do not necessarily have to meet one of A, B, C. As they still cannot have meet multiple of A, B, C, there are now $1 + 3 = 4$ choices each for D and E (including a choice for none). However, this overcounts the 3 outcomes where D and E met the same person (forming a second triangle). Thus there are $4 \cdot 4 - 3 = 13$ outcomes for this case.

Putting everything together we have a final answer of

$$\binom{5}{3}(3\cdot 3 + 4\cdot 4 - 3) = 10\cdot 22 = 220.$$

Answer: 220

2.2   ZIML November 2018 Junior Varsity

Below are the solutions from the Junior Varsity ZIML Competition held in November 2018.

The problems from the contest are available on p.25.

Problem 1 Solution

Since $8 = 2^3$ each digit in the octal number correspond to 3 digits of the binary number. Further, we can do this conversion digit by digit. We have

$$3 = 11_2$$
$$1 = 1_2 = 001_2$$
$$2 = 10_2 = 010_2$$
$$3 = 11_2 = 011_2$$

Therefore $3123_8 = 11001010011_2$ so our digits are 11001010011.

Answer: 11001010011

Problem 2 Solution

We start by arranging the girls, which can be done in $5! = 120$ ways. This creates 6 spaces for the boys to stand in the photo.

As 2 of the boys stand together, but not all 3, we pick one of the spaces for 2 boys and a different space for the third boy. Hence there are $6 \cdot 5 = 30$ ways to choose these spaces.

For the space with 2 boys, there are $3 \cdot 2 = 6$ ways to choose and order two of the boys, and the third boy stands in the remaining spot.

Hence there are

$$5! \cdot 6 \cdot 5 \cdot 3 \cdot 2 = 120 \cdot 30 \cdot 6 = 21600$$

arrangements in total.

Answer: 21600

Problem 3 Solution

$\widehat{BC} = 160° - 60° = 100°$. As a full circle measures $360°$,

$$\widehat{CD} = 360° - 60° - 100° - 140° = 60°,$$

so $\widehat{AB} = \widehat{CD}$. Using inscribed angles we have

$$\angle ABC = \frac{1}{2} \cdot (\widehat{CD} + \widehat{AD}) = \frac{1}{2} \cdot (\widehat{AB} + \widehat{AD}) = \angle BCD.$$

An identical argument gives $\angle BAD = \angle CDA$. Therefore $ABCD$ is a trapezoid.

It follows that $[ABE] = [CDE]$. ($\triangle ABD$ and $\triangle ACD$ share the same height, hence same area, and subtracting $\triangle AED$ gives the result.) Thus the ratio $\dfrac{[ABE]}{[CDE]} = \dfrac{1}{1} = 1$.

Answer: 1

Problem 4 Solution

We let $u = x^2 + 5x - 3$. Note subtracting 6 from both sides we get, after substituting,

$$u^3 + 2u = u^2 \text{ or } u(u^2 - u + 2) = 0.$$

Note $u^2 - u + 2$ has no real solutions as the discriminant is $1 - 8 = -7$. Hence we are left with

$$u = x^2 + 5x - 3 = 0.$$

Using the discriminant again, $25 - (-12) = 37$, so this has 2 real solutions.

Therefore our original equation has 2 real solutions.

Answer: 2

Problem 5 Solution

You want the probability of getting at least one green ball to be $\geq 80\%$, which is equivalent to the probability of getting two red balls being $< 20\%$.

With 10 red balls, there are $\binom{10}{2} = 45$ ways to choose 2 red balls. Similarly, with $10 + K$ balls in total, there are $\binom{10+K}{2} = \frac{(10+K)(9+K)}{2}$ ways to choose 2 balls. Thus we want

$$45 \div \frac{(10+K)(9+K)}{2} = \frac{90}{(10+K)(9+K)} < 20\% = \frac{1}{5}.$$

Cross multiplying we have $450 < (10+K)(9+K)$. Noting $21^2 = 441$ and $22 \cdot 21 = 462$ it is clear that $K = 22 - 10 = 12$ is the smallest value of K that works.

(Alternatively we can solve the quadratic inequality for K, but here we take advantage of the fact that K is an integer and $(10 + K)(9 + K)$ is approximately a square.)

Answer: 12

Problem 6 Solution

Our inequality can be rewritten as $5x^2 + y^2 - 4xy - 6x + 8 \leq 0$. Considering the terms with y: $y^2 - 4xy$ we see that including a $4x^2$ to complete the square we have $y^2 - 4xy + 4x^2 = (y - 2x)^2$. Hence we have

$$x^2 - 6x + 8 + (y - 2x)^2 \leq 0.$$

Completing the square once more we have

$$(x - 3)^2 + (y - 2x)^2 \leq 1.$$

If we let $A = x - 3$ and $B = y - 2x$ we have $A^2 + B^2 \leq 1$. As x, y, A, and B are integers, we consider 3 cases: (i) $A = B = 0$, (ii) $A = 1, B = 0$, or (iii) $A = 0, B = 1$.

In case (i), $x = 3$ so $y = 2 \cdot 3 = 6$, giving one pair.

In case (ii), x is either 2 or 4 meaning y is one of $2 \cdot 2 = 4$ or $2 \cdot 4 = 8$ giving two additional pairs.

In case (iii), $x = 3$ again, but y is either $1 + 2 \cdot 3 = 7$ or $-1 + 2 \cdot 3 = 5$ giving two final pairs.

Hence there are $1 + 2 + 2 = 5$ integer pair solutions in total.

Answer: 5

Problem 7 Solution
The inscribed circle's center is the incenter, the intersection of the angle bisectors in $\triangle ABC$. This gives the diagram below:

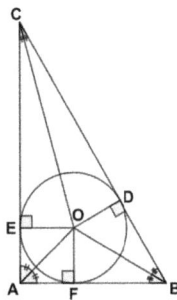

$\angle FBO = 60° \div 2 = 30°$ so $\triangle BFO$ is a 30-60-90 triangle. We know $OD = OE = OF = 3$ so $\triangle BFO$ and hence $BF = OF \cdot \sqrt{3} = 3\sqrt{3}$. Hence

$$AF = AB - BF = (3\sqrt{3} + 3) - 3\sqrt{3} = 3.$$

This means $\triangle AFO$ is an isosceles right triangle ($AF = OF = 3$) so $\angle FAO = 45°$.

Therefore $\angle BAC = 90°$ so in fact $\triangle ABC$ is a 30-60-90 triangle. Thus

$$AC = AB \cdot \sqrt{3} = (3\sqrt{3} + 3) \cdot \sqrt{3} = 9 + 3\sqrt{3}.$$

Rewriting we have $AC = \sqrt{81} + \sqrt{27}$ so $K + L = 27 + 81 = 108$.

Answer: 108

Problem 8 Solution
Factoring we have $204020 = 2020 \cdot 101 = 2^2 \cdot 5 \cdot 101^2$. Clearly 2020 is a factor larger than 2018 and because factors come in pairs, any other factors larger than 2020 will have a corresponding factor less than $204020 \div 2020 = 101$.

Looking at the prime factorization ($2^2 \cdot 5 \cdot 101^2$), these all must be factors of $2^2 \cdot 5$, so there are $(2 + 1)(1 + 1) = 6$ such factors.

Hence we have $6 + 1 = 7$ factors in total.

Answer: 7

Problem 9 Solution
We consider $x < -2$ and $x \geq -2$ as two separate cases.

If $x < -2$ then $|x + 2| = -(x + 2) = -x - 2$. Our equation is thus $\sqrt{x^2 + 2} = -x$. Squaring both sides we have $x^2 + 2 = x^2$ so there are no solutions in this case.

If $x \geq -2$ then $|x + 2| = x + 2$. Our equation is thus $\sqrt{x^2 + 2} = x + 4$. Squaring both sides we have $x^2 + 2 = x^2 + 8x + 16$. Simplifying we have $8x = -14$ so $x = -\dfrac{7}{4} = -1.75$.

Thus $x = -1.75$ is the only real solution, so our answer is -1.75.

Answer: -1.75

Problem 10 Solution

Consider 3 cases: (i) the point is chosen inside the circle of radius 1, (ii) outside the circle of radius 1 but in the circle of radius 3, or (ii) outside the circle of radius 2 (but inside the circle of radius 3).

The three circles have areas $\pi \cdot 1^2 = \pi$, $\pi \cdot 2^2 = 4\pi$, and $\pi \cdot 3^2 = 9\pi$. Hence the three cases have respective probabilities of

$$\frac{\pi}{9\pi} = \frac{1}{9}, \quad \frac{4\pi - \pi}{9\pi} = \frac{1}{3}, \quad \text{and} \quad \frac{9\pi - 4\pi}{9\pi} = \frac{5}{9}.$$

In case (i), there are 3 gray regions (our of 5 total) to the probability of the point being in a gray region is $\frac{3}{5}$. Similar reasoning gives probabilities of $\frac{2}{5}$ for case (ii) and $\frac{3}{5}$ for case (iii).

Hence using the law of total probability, the point being in a gray region has probability

$$\frac{1}{9} \cdot \frac{3}{5} + \frac{1}{3} \cdot \frac{2}{5} + \frac{5}{9} \cdot \frac{3}{5} = \frac{8}{15}$$

so $R + S = 23$.

Answer: 23

Problem 11 Solution

Since the point is on $y = -x$, we have the point can be written as $(m, -m)$ for an integer m.

The distance from $(m, -m)$ to $(0, 4)$ is

$$\sqrt{(m-0)^2 + (-m-4)^2} = \sqrt{2m^2 + 8m + 16}$$

using the distance formula.

Note the lines $y = -x$ and $y = x$ are perpendicular. Thus the shortest distance from $(m, -m)$ to $y = x$ is from $(m, -m)$ to the intersection point $(0, 0)$. Hence this distance is $m\sqrt{2}$.

Hence we want

$$\sqrt{2m^2 + 8m + 16} = m\sqrt{2}$$
$$\Rightarrow 2m^2 + 8m + 16 = 2m^2$$
$$\Rightarrow m = -2$$

Thus the point is $(-2, 2)$ so $m \times n = -2 \times 2 = -4$.

Answer: -4

Problem 12 Solution

We work modulo 31. By Fermat's Little Theorem, $11^{30} \equiv 1 \pmod{31}$. Thus

$$11^{2018} \equiv 11^{2010} \cdot 11^8 = 1 \cdot 11^8 \pmod{31},$$

as 2010 is a multiple of 30. We have

$$11^8 \equiv (11^2)^4 \equiv 121^4 \equiv (-3)^4 \pmod{31}$$
$$\equiv 81 \equiv 19 \pmod{31}.$$

Thus the remainder when 11^{2018} is divided by 31 is 19.

Answer: 19

Problem 13 Solution

Note substituting $d = a + b$ and $e = c + d$ into the third equation we have

$$c + d + e = c + d + c + d = 2c + 2d = 2a + 2b + 2c = 18.$$

Further, a, b, and c uniquely determine d and e.

Hence it is enough to count solutions to

$$2a + 2b + 2c = 18 \text{ or } a+b+c = 9$$

where a, b, and c are non-negative integers. Using stars and bars there are

$$\binom{9+3-1}{9} = \frac{11}{9} = 55$$

such solutions.

Answer: 55

Problem 14 Solution

Expanding and simplifying we have

$$x^2 + (y-10)^2 = \frac{1}{10}(y-3x)^2$$
$$10x^2 + 10y^2 - 200y + 1000 = y^2 - 6xy + 9x^2$$
$$x^2 + 6xy + 9y^2 - 200y + 1000 = 0$$

Consider this as a quadratic in x (as if (x,y) is a solution, then the quadratic in x must have real solutions for x). Using the discriminant there are real solutions for x when

$$(6y)^2 - 4(9y^2 - 200y + 1000) \geq 0$$
$$800y - 4000 \geq 0$$

so $y \geq 5$. Thus $K = 5$.

(One can calculate that the point $(-15,5)$ is the minimum of all solutions to our equation.)

Answer: 5

Problem 15 Solution

We first calculate the lcm$(1332, 2376)$. Using prime factorizations we have

$$1332 = 2^2 \cdot 3^2 \cdot 37 \text{ and } 2376 = 2^3 \cdot 3^3 \cdot 11$$

and hence lcm$(1332, 2376) = 2^3 \cdot 3^3 \cdot 11 \cdot 37$.

N must be a multiple of this LCM and since it has an odd number of factors, must be a perfect square. Hence all the exponents are even and thus

$$N = 2^4 \cdot 3^4 \cdot 11^2 \cdot 37^2.$$

N therefore has $(4+1)(4+1)(2+1)(2+1) = 225$ factors.

Answer: 225

Problem 16 Solution

Consider a side view of the operative's path, giving the right triangle below (where S is the starting point, E is the ending point, and R denotes the radar station).

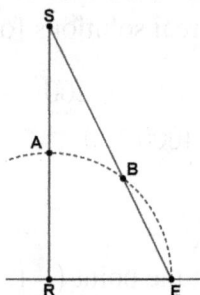

If A and B denote, respectively, the intersections of SR and SE with the circle, we have

$$SR = 2, RA = AS = 1, RE = 1 \Rightarrow SE = \sqrt{1^2 + 2^2} = \sqrt{5}$$

Let C be such that AC is a diameter. Using power of a point,

$$SA \cdot SC = SB \cdot SE$$
$$\Rightarrow 1 \cdot 3 = SB \cdot \sqrt{5}$$
$$\Rightarrow SB = \frac{3}{\sqrt{5}} = \frac{3\sqrt{5}}{5}$$
$$\Rightarrow BE = \sqrt{5} - \frac{3\sqrt{5}}{5} = \frac{2\sqrt{5}}{5}$$

Therefore,
$$\frac{2\sqrt{5}}{5} \div \sqrt{5} = \frac{2}{5} = 40\%$$

of the operative's path is within 1 mile of the radar station and hence $K = 40$.

Answer: 40

Problem 17 Solution

We first observe that in fact

$$4a^2 + b^2 + c^2 + 4ab + 4ac + 2bc = (2a+b+c)^2.$$

Using Vieta's theorem, we know that $a+b+c = -\frac{-10}{2} = 5$. Thus

$$(2a+b+c)^2 = [a+(a+b+c)]^2 = [3+5]^2 = 8^2 = 64,$$

giving an answer of 64.

Answer: 64

Problem 18 Solution

We use the binomial theorem to help evaluate the sum.

Note $64 = 2^6$ and $729 = 3^6$. Checking, $96 = 2^5 \cdot 3$, $144 = 2^4 \cdot 3^2$, etc. Thus our sum is

$$(2-3)^6 = (-1)^6 = 1$$

using the binomial theorem.

Answer: 1

Problem 19 Solution

We are looking for the ratio $[EFG] : [E'F'G']$ (here $[-]$ denotes area). Consider the example diagram below.

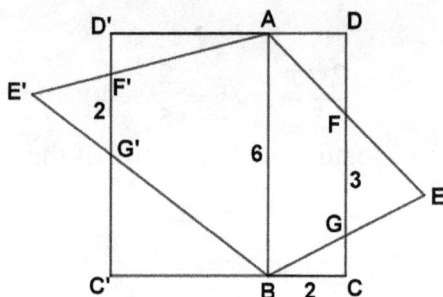

We know $ABCD$ is a 6×2 rectangle while $ABC'D'$ is a 6×4 rectangle. Both $ABGF$ and $ABG'F'$ are trapezoids with areas

$$[ABGF] = \frac{AB + FG}{2} \cdot AD = \frac{6+3}{2} \cdot 2 = 9$$

$$[ABG'F'] = \frac{AB + F'G'}{2} \cdot AD' = \frac{6+2}{2} \cdot 4 = 16$$

Now $\triangle ABE \sim \triangle FGE$ with ratio of sides $6 : 3 = 2 : 1$. Hence

$$[ABE] : [FGE] = 4 : 1 \Rightarrow [ABGF] : [FGE] = 3 : 1.$$

Thus $[FGE] = \frac{1}{3} \cdot 9 = 3$.

Similarly $\triangle ABE' \sim \triangle F'G'E'$ with ratio of sides $3 : 1$ so

$$[ABG'F'] : [F'G'E'] = 8 : 1.$$

Thus $[F'G'E'] = \dfrac{1}{8} \cdot 16 = 2$.

Hence $[EFG] : [E'F'G'] = 3 : 2$ and $P + Q = 3 + 2 = 5$.

Answer: 5

Problem 20 Solution

$88 = 8 \cdot 11$ so we want our number to be divisible by 8 and 11.

To be divisible by 8 we need the last 3 digits to be divisible by 8, so $\overline{c16}$ to be divisible by 8. Thus we have $c = 0, 2, 4, 6$, or 8.

To be divisible by 11 we need the alternating sum of the digits

$$6 - 1 + c - b + a - 0 + 2 = a - b + c + 7$$

to be a multiple of 11. Hence we want

$$a - b + c + 7 \equiv 0 \pmod{11} \text{ or } a \equiv b - c + 4 \pmod{11}.$$

Thus for any b and c, we can find a unique a such that $0 \le a < 11$ and hence we initially have 10 pairs of a and b for each of the values of c above. However, note that $a = 10$ is not a valid digit. Unless $c = 4$ (so $a \equiv b \pmod{11}$), this removes 1 pair of a and b for each c value.

Hence there are $9 + 9 + 10 + 9 + 9 = 46$ seven digit numbers satisfying the needed restrictions.

Answer: 46

2.3   ZIML December 2018 Junior Varsity

Below are the solutions from the Junior Varsity ZIML Competition held in December 2018.

The problems from the contest are available on p.33.

Problem 1 Solution

We want to separate the sixth graders, so arrange the seventh graders and 2 teachers first. The seventh graders can be arranged in 3! ways, with 2! ways to arrange the teachers outside the seventh graders.

This creates $5 + 1 = 6$ spaces for the sixth graders. Hence there are $6 \cdot 5 \cdot 4$ ways to place the 3 sixth graders.

Hence there are
$$3! \cdot 2! \cdot 6 \cdot 5 \cdot 4 = 1440$$
arrangements and hence 1440 different possible photographs.

Answer: 1440

Problem 2 Solution

Note that $(20, 21, 29)$ is a Pythagorean triple, so $\angle ABC = 90°$ and AC is the diameter of the circle. Thus $\angle ADC = 90°$ as well, so $\widehat{AD} = \widehat{DC} = 180° \div 2 = 90°$ and hence $\triangle ADC$ is a 45-45-90 triangle. This gives the diagram below.

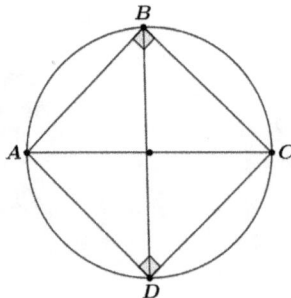

Thus $AD = CD = \dfrac{29}{\sqrt{2}} = \dfrac{29\sqrt{2}}{2}$. Using Ptolemy's theorem we have

$$AC \cdot BD = AB \cdot CD + BC \cdot AD$$

$$29 \cdot BD = 20 \cdot \frac{29\sqrt{2}}{2} + 21 \cdot \frac{29\sqrt{2}}{2}$$

$$29 \cdot BD = \frac{29\sqrt{2}}{2} \cdot 41$$

$$BD = \frac{41\sqrt{2}}{2}.$$

Therefore $R + S + T = 41 + 2 + 2 = 45$.

Answer: 45

Problem 3 Solution

To be divisible by $33 = 3 \times 11$ the number must be divisible by 3 and by 11. Hence the sum of the digits must be divisible by 3 and the alternating sum of the digits must be divisible by 11.

Consider first a three digit number of the form $\overline{ABC}$ (the digits are A, B, and C). We thus want

$$A + B + C \equiv 0 \pmod{3} \text{ and } C - B + A \equiv 0 \pmod{11}.$$

Since A, B, and C are digits, the only way $A + C - B \equiv 0 \pmod{11}$ is if $A + C - B = 0$ and hence $A + C = B$. However, this implies $A + B + C = 2B$ which is even, so not divisible by 3.

Therefore no three digit numbers work. Considering now a number $\overline{ABCD}$ we need

$$A + B + C + D \equiv 0 \pmod{3} \text{ and } D - C + B - A \equiv 0 \pmod{11}.$$

Clearly any digits $A = B$, $C = D$ give a number divisible by 11. For $A + B + C + D \equiv 0 \pmod{3}$, we see the smallest such is $A = B = 1$ and $C = D = 2$. Therefore the number 1122 works.

Double checking, no other 4 digit multiple of 33 works, so 1122 is the smallest number.

Answer: 1122

Problem 4 Solution
First note $x^2 - 3x - 4 = (x-3)(x+1)$ so $x \neq 3$ and $x \neq -1$. Using polynomial long division we have

$$(x^4 - 5x^3 - x^2 + 17x + 12) \div (x^2 - 3x - 4) = x^2 - 2x - 3.$$

Hence solving $x^2 - 2x - 3 = 0$ we get $(x-3)(x+1) = 0$ so $x = 3$ or $x = -1$. However, $x = -1$ is not in the domain, so $x = 3$ is the only real solution. Hence our answer is 3.

Answer: 3

Problem 5 Solution
Note calculating in base 8 we have (identical to base 10)

$$3456_8 - 2345_8 = 1111_8.$$

Converting this to base 10 we have

$$1111_8 = 8^3 + 8^2 + 8 + 1 = 585$$

as our answer.

Answer: 585

Problem 6 Solution
Each pizza has 3 toppings, so (as repeated toppings are allowed) there are $\binom{3+8-1}{3} = \binom{10}{3} = 120$ different pizzas Jerry can choose from. Since Jerry wants a different pizza each day, there are $120 \cdot 119 = 14280$ ways for Jerry to order pizza this weekend.

Answer: 14280

Problem 7 Solution

AD is the angle bisector of $\angle BAC$, so using the angle bisector theorem we have

$$\frac{BD}{CD} = \frac{AB}{AC} = \frac{15}{12} = \frac{5}{4}.$$

Write the area of $\triangle ABC$ as $[ABC] = 72$. Since $\triangle BCE$ and $\triangle ABC$ share the same height and E is the midpoint of AB we have

$$[BCE] = \frac{1}{2} \cdot [ABC] = \frac{1}{2} \cdot 72 = 36.$$

Similarly, as $\triangle BDE$ and $\triangle BCE$ share the same height and $\dfrac{BD}{BC} = \dfrac{5}{9}$, we have

$$[BDE] = \frac{5}{9} \cdot [BCE] = \frac{5}{9} \cdot 36 = 20.$$

Hence our answer is 20.

Answer: 20

Problem 8 Solution

N is divisible by $126 = 2 \cdot 3^3 \cdot 7$, so we already know it has 3 different prime factors. As 5 is the smallest prime not included, N should also be a multiple of 5.

Thus N should be the smallest multiple of $2 \cdot 3^3 \cdot 5 \cdot 7$ that is a perfect square. This means all the exponents must be even, so $N = 2^2 \cdot 3^4 \cdot 5^2 \cdot 7^2$. Therefore N has $(2+1)(4+1)(2+1)(2+1) = 135$ factors.

Answer: 135

Problem 9 Solution

Let integers a, b, c, d, and e be roots of $Q(x)$ such that a, b, c are roots of $P(x)$.

Using Vieta's theorem for $P(x)$ we have that

$$a+b+c = -\frac{-1}{1} = 1$$
$$\text{and } a \cdot b \cdot c = -\frac{-8}{1} = 8.$$

Similarly for $Q(x)$ we have

$$a+b+c+d+e = -\frac{4}{1} = -4$$
$$\text{and } a \cdot b \cdot c \cdot d \cdot e = -\frac{192}{1} = -192.$$

Therefore, subtracting and dividing these equations respectively gives
$$d+e = -5 \text{ and } d \cdot e = -24.$$

Solving this system we have $d, e = -8, 3$.

As a, b, c are all integers with $a \cdot b \cdot c = 8$, we must have $x = -8$ is the smallest root of $Q(x)$.

(Note: Using the fact that all the roots are integers, it is possible to determine that the 5 roots are actually -8, -2, -1, 3, and 4.)

Answer: -8

Problem 10 Solution

Reflect B across the line $y = x$ to a point B' as shown in the diagram below.

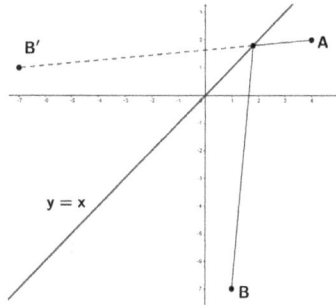

Then (as in the diagram) the shortest path from A to B which touches the line $y = x$ can be determined by reflecting back across $y = x$ the straight line from A to B'.

Therefore the length of the shortest path is just the distance from A to B'. Reflecting across the line $y = x$ just swaps x and y-coordinates, so $B' = (-7, 1)$. Hence this distance is

$$\sqrt{(4 - (-7))^1 + (2 - 1)^2} = \sqrt{121 + 1} = \sqrt{122}$$

giving a final answer of $D = 122$.

Answer: 122

Problem 11 Solution

There are 6^4 total outcomes.

There are 3 primes (2, 3, and 5) and 2 composites (4 and 6) among. Hence there are $3^2 \cdot 2 \cdot 1$ ways to determine the numbers rolled. We also, however, need to determine their order. As there are two primes, one composite, and one 1, there are $\binom{4}{2} \cdot 2!$ orderings.

Thus the probability is

$$\frac{\binom{4}{2} \cdot 2! \cdot 3^2 \cdot 2 \cdot 1}{6^4} = \frac{6^3}{6^4} = \frac{1}{6}$$

so $Q - P = 6 - 1 = 5$.

Answer: 5

Problem 12 Solution

Consider this as a recursive sequence $x_0, x_1, x_2, \ldots$ with

$$x_0 = S, x_{n+1} \equiv 11 \cdot x_n + 5.$$

Therefore

$$
\begin{aligned}
x_{n+2} &\equiv 11 \cdot x_{n+1} + 5 \quad (\text{mod } 100) \\
&\equiv 11 \cdot (11 \cdot x_n + 5) + 5 \quad (\text{mod } 100) \\
&\equiv 121 \cdot x_n + 60 \quad (\text{mod } 100) \\
&\equiv 21 \cdot x_n + 60 \quad (\text{mod } 100)
\end{aligned}
$$

If the sequence alternates between two terms (starting with S) we must have

$$
\begin{aligned}
S &\equiv 21S + 60 \quad (\text{mod } 100) \\
\Rightarrow 20S &\equiv -60 \quad (\text{mod } 100) \\
\Rightarrow 20S &= -60, 40, 140, 240, \ldots \\
\Rightarrow S &= -3, 2, 7, 12, \ldots
\end{aligned}
$$

Thus $S = 5k - 3$ will work for $k = 1, 2, \ldots, 20$. Hence there are 20 initial numbers S.

Answer: 20

Problem 13 Solution

Assume both polygons have side length 1, so the area of the square is 1. Thus the ratio of areas is just the area of the octagon.

The external angles of a regular octagon are all $45°$, so we can extend opposite sides of the octagon to create a square, made up of the octagon plus four $45 - 45 - 90$ right triangles. Each of these triangles has hypotenuse of 1, so legs of length $\dfrac{1}{\sqrt{2}} = \dfrac{\sqrt{2}}{2}$.

Therefore we can calculate the area of the octagon as the area of this square minus the four right triangles:

$$(1+\sqrt{2})^2 - 4 \cdot \frac{1}{4} = 2 + 2\sqrt{2},$$

which is M. Since $2\sqrt{2} = \sqrt{8}$, we have $\lfloor 2\sqrt{2} \rfloor = 2$. Hence $\lfloor M \rfloor = \lfloor 2 + 2\sqrt{2} \rfloor = 4$.

Answer: 4

Problem 14 Solution
We know $G_n = 2 \cdot F_n - 2$. Plugging in $F_n = F_{n-1} + F_{n-2}$ we have

$$\begin{aligned} G_n &= 2 \cdot F_{n-1} + 2 \cdot F_{n-2} - 2 \\ &= 2 \cdot F_{n-1} - 2 + 2 \cdot F_{n-2} + 2 \\ &= G_{n-1} + G_{n-2} + 2 \end{aligned}$$

Therefore $a = b = 1$ and $c = 2$ so $a + b + c = 1 + 1 + 2 = 4$.

Answer: 4

Problem 15 Solution
We have that $45 = 5 \cdot 9$ so by the Chinese Remainder Theorem it is enough to calculate $2019^{2019} \pmod 5$ and $2019^{2019} \pmod 9$ to recover the answer.

For 5, we have that

$$2019^{2019} \equiv (-1)^{2019} \equiv -1 \equiv 4 \pmod 5.$$

For 9 first note $2019 \equiv 2 + 0 + 1 + 9 \equiv 3 \pmod 9$. Therefore

$$2019^{2019} \equiv 3^{2019} \equiv 0 \pmod 9.$$

Hence the remainder r is the number $0 \leq r < 45$ so that $r \equiv 4$ (mod 5) and $r \equiv 0$ (mod 9). Thus $r = 9$ is our answer.

Answer: 9

Problem 16 Solution

The perpendicular bisector of $\overline{AB}$ is all points that are equidistant from A to B, so the points in the shaded region below are closer to A than to B.

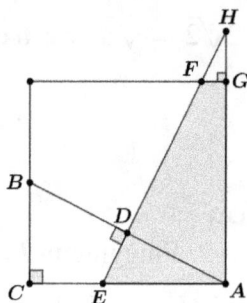

(Here D is the midpoint of $\overline{AB}$, so $\overleftrightarrow{EF}$ is the perpendicular bisector.)

Suppose the square has a side length of 2, so $AC = 2$, $BC = 1$, and thus $AB = \sqrt{2^2 + 1^2} = \sqrt{5}$. Using that $\triangle ADE \sim \triangle ACB$,

$$\frac{AE}{AD} = \frac{AB}{AC} \Rightarrow AE = \frac{AB \cdot AD}{AC}$$

$$\Rightarrow AE = \sqrt{5} \cdot \frac{\sqrt{5}}{2} \div 2$$

$$\Rightarrow AE = \frac{5}{4}$$

If H is the intersection of the perpendicular bisector and the extension of the side of the square (as in the diagram above), then we also have that $\triangle EAH \sim \triangle BCA \sim \triangle FGH$. As $AC:BC=2:1$ we have $AH = 2 \cdot AE = 2 \cdot \dfrac{5}{4} = \dfrac{5}{2}$. Hence

$$GH = AH - AG = \frac{5}{2} - 2 = \frac{1}{2}$$

and then

$$FG = GH \div 2 = \frac{1}{2} \div 2 = \frac{1}{4}.$$

Therefore $AEFG$ (a trapezoid) has area

$$\frac{1}{2} \cdot (AE + GF) \cdot AG = \frac{1}{2} \cdot \left(\frac{5}{4} + \frac{1}{4} \right) \cdot 2 = \frac{3}{2}.$$

As the full square has area 4, our probability is

$$\frac{3}{2} \div 4 = \frac{3}{8} = 0.375 = 37.5\%$$

and hence $K = 37.5$.

Answer: 37.5

Problem 17 Solution

Completing the square inside each radical we have

$$6x - x^2 = -(x^2 - 6x)$$
$$= -(x^2 - 6x + 9) + 9$$
$$= -(x - 3)^2 + 9$$

and similarly

$$4x^2 - 24x + 52 = 4(x^2 - 6x + 13)$$
$$= 4(x^2 - 6x + 9) + 16$$
$$= 4(x - 3)^2 + 16.$$

Therefore

$$f(x) = \sqrt{9 - (x-3)^2} - \sqrt{4(x-3)^2 + 16}.$$

As $(x-3)^2 \geq 0$ for all x, we see that $x = 3$ will maximize the first radical and minimize the second radical. Therefore the maximum value of $f(x)$ is

$$f(3) = \sqrt{9} - \sqrt{16} = 3 - 4 = -1.$$

Hence $f(x) \leq -1$ for all x, so $B = -1$.

Answer: -1

Problem 18 Solution

Factoring we have

$$|(x+3)^2| + 6 = |5(x+3)|$$

or equivalently $|x+3|^2 + 6 = 5|x+3|$.

Hence after the substitution $y = |x+3|$ we have

$$y^2 - 5y + 6 = (y-2)(y-3) = 0$$

so $y = 2$ or $y = 3$. Solving we have $|x+3| = 2$ so $x = -1, -5$ or $|x+3| = 3$ so $x = 0, -6$. Double checking, $x = -6$, $x = -5$, $x = -1$, and $x = 0$ are all solutions, with sum -12.

Answer: -12

Problem 19 Solution

The original pyramid is cut into two pieces. The bottom piece is shown in the problem.

Since the plane cutting the pyramid is parallel to the base, the other piece is a smaller right square pyramid, similar to the first. The height of the smaller pyramid is half that of the original pyramid, so its base area and surface area are respectively

$$\left(\frac{1}{2}\right)^2 \cdot 36 = 9 \text{ and } \left(\frac{1}{2}\right)^2 \cdot 96 = 24.$$

Then we note that the surface area of the larger piece shown in the problem is equal to the surface area of the full pyramid minus the smaller similar pyramid, plus twice the base of the smaller pyramid. (This base is the top of the larger piece, and must be included.) Hence the surface area of the larger piece is

$$96 - 24 + 2 \cdot 9 = 90,$$

which is our answer.

Answer: 90

Problem 20 Solution
By factoring, $x^2 - 3x - 4 = (x - 4)(x + 1)$ so $x^2 - 3x - 4$ is negative between -1 and 4. Furthermore, $L = 1$ and $L = -4$ clearly work, as then the graphs intersect when $y = 0$. Hence we need to consider cases when $x < -1$ or $x > 4$.

Directly solving we have the graphs intersect when

$$x^2 - 3x - 4 = x^2 + Lx \Rightarrow x = -\frac{4}{L + 3},$$

so we are left to ensure $x < -1$ or $x > 4$. The graphs do not intersect when $L = -3$ (as we are dividing by 0). When $L < -3$, $0 < x \leq 4$, so we focus on $L > -3$. In this case, $x < 0$. For $x < -1$ we need $L + 3 < 4$ so $L < 1$.

Hence the graphs intersect at a positive y value when $L = -2, -1, 0$ and at $y = 0$ when $L = -4, 1$. This gives 5 integers L.

Answer: 5

2.4 ZIML January 2019 Junior Varsity

Below are the solutions from the Junior Varsity ZIML Competition held in January 2019.
The problems from the contest are available on p.41.

Problem 1 Solution

Calculate the values 1^2, 2^2, 3^2, ..., 9^2, 10^2, and we find that only $4^2 = 16$ and $6^2 = 36$ have odd numbers as the tens digits.

Let $n = 10a + b$, where a and b are digits, then

$$n^2 = 100a^2 + 20ab + b^2 = 20a(5a + b) + b^2,$$

and its tens digits equals an even number plus the tens digit of b^2. Therefore the tens digit of n^2 is an odd number if and only if $b = 4$ or $b = 6$. Let $a = 0, 1, 2, \ldots 9$, then there are $10 \times 2 = 20$ such numbers whose squares have odd numbers as the tens digits.

Answer: 20

Problem 2 Solution

From $x^2 - 11x + 1 = 0$, we get

$$x^2 + 1 = 11x,$$

thus

$$x + \frac{1}{x} = 11,$$

squaring both sides,

$$x^2 + 2 + x^{-2} = 121,$$

which means

$$x^2 + x^{-2} = 119,$$

and squaring both sides again,

$$x^4 + 2 + x^{-4} = 14161,$$

therefore

$$x^4 + x^{-4} = 14159.$$

Answer: 14159

Problem 3 Solution

First we have $a_5 = a_4 + \binom{4}{2} = 10 + \binom{4}{2}$. Similarly we have

$$a_6 = a_5 + \binom{5}{3} = 10 + \binom{4}{2} + \binom{5}{3}.$$

Continuing this pattern we see

$$a_{20} = 10 + \binom{4}{2} + \binom{5}{3} + \cdots + \binom{19}{17}.$$

By the symmetry $\binom{n}{k} = \binom{n}{n-k}$,

$$a_{20} = 10 + \binom{4}{2} + \binom{5}{2} + \cdots + \binom{19}{2}.$$

By the hockey stick identity,

$$\binom{20}{3} = \binom{2}{2} + \binom{3}{2} + \binom{4}{2} + \cdots + \binom{19}{2}$$
$$= \binom{2}{2} + \binom{3}{2} + a_{20} - 10$$
$$= a_{20} - 6.$$

Therefore $a_{20} = \binom{20}{3} + 6 = 1140 + 6 = 1146.$

Answer: 1146

Problem 4 Solution

Let $d = \gcd(a,b)$, then $a = md$ and $b = nd$, where $m > n$ and $\gcd(m,n) = 1$. Thus $\text{lcm}(a,b) = mnd$, so

$$md - nd = 120, \quad mnd = 105d,$$

hence

$$mn = 105.$$

Since $m > n$, there are 4 possibilities: $(m,n) = (105,1)$, $(35,3)$, $(21,5)$, or $(15,7)$. Since $md - nd = 120$, $m - n$ must be a factor of 120, and only $(15,7)$ satisfies this requirement. Therefore $m = 15, n = 7, d = 15$, and then $a = md = 225$, $b = nd = 105$, and $a + b = 330$.

Answer: 330

Problem 5 Solution

Since

$$\alpha = \min\{\angle A - \angle B, \angle B - \angle C, 90° - \angle A\},$$

we have

$$\alpha \leq \angle A - \angle B, \alpha \leq \angle B - \angle C, \alpha \leq 90° - \angle A.$$

Therefore,

$$\begin{aligned} 6\alpha &\leq 2(\angle A - \angle b) + (\angle B - \angle C) + 3(90° - \angle A) \\ &= 270° - (\angle A + \angle B + \angle C) \\ &= 90°, \end{aligned}$$

therefore $\alpha \leq 15°$. The maximum value $15°$ is attained when $\angle A = 75°$, $\angle B = 60°$, and $\angle C = 45°$.

Answer: 15

Problem 6 Solution

Since you can use each edge and diagonal at most once, it is impossible to visit only one other vertex.

Given any two vertices, there is exactly one line segment (edge or diagonal) connecting them. Thus, counting paths is equivalent to counting permutations of 2 or more of B, C, D, and E. For example, permuting B, C, and E could give the path

$$A \to C \to B \to E \to A.$$

Considering cases of 2, 3, or 4 other vertices, we see there are

$$\frac{4!}{2!} + \frac{4!}{1!} + \frac{4!}{0!}$$
$$= 12 + 24 + 24$$
$$= 60$$

total paths.

Answer: 60

Problem 7 Solution

Label Jackie's three jumps x, y, and z and consider them plotted in three dimensions.

We know $0 \le x, y, z \le 2$ so all possible outcomes form a cube with side length 2 and hence volume 8.

We want the sum of Jackie's jumps to be at least 4 meters, so $x + y + z = 4$. This intersects the cube at the vertices $(2,2,0)$, $(2,0,2)$, and $(0,2,2)$ so the region we want is the triangular pyramid shown below.

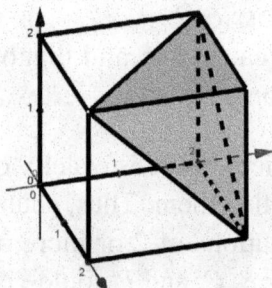

This pyramid has volume

$$\frac{1}{3} \cdot \frac{2^2}{2} \cdot 2 = \frac{4}{3}.$$

Therefore the probability Jackie achieves her goal is

$$\frac{4}{3} \div 8 = \frac{1}{6} = 0.1\overline{6} = 16.\overline{6}\%.$$

Therefore rounded to the nearest integer, $K = 17$.

Answer: 17

Problem 8 Solution

Let $x = \overline{ab}$ and $y = \overline{cd}$, then $10 \le x \le 99$ and $0 \le y \le 99$. Note that x is at least 10 and y is allowed to be 0 or a one-digit number. Then

$$100x + y = (x+y)^2.$$

Expand and write as a quadratic equation in x,

$$x^2 + 2(y - 50)x + (y^2 - y) = 0.$$

Since the equation has real solutions, the discriminant is nonnegative, that is

$$4(y - 50)^2 - 4(y^2 - y) = 4(2500 - 99y) \ge 0,$$

therefore $y \leq 25$, and

$$x = (50 - y) \pm \sqrt{2500 - 99y}.$$

Since x and y are both integers, $2500 - 99y$ must be a perfect square. Let k be a positive integer such that

$$2500 - 99y = k^2,$$

then

$$99y = (50 + k)(50 - k),$$

then either $50 + k$ or $50 - k$ is a multiple of 11, and so $k = 49$, 38, 27, 16, 5, 6, 17, 28, and 39, Trying these values, only $k = 49$ and $k = 5$ give integer values for y. If $k = 49$, $y = 1$ and $x = 98$ (another solution $x = 0$ is not valid). If $k = 5$, $y = 25$ and $x = 20$ or 30.

Therefore there are three possible solutions: $N = 9801$, $N = 2025$ and $N = 3025$. The sum is $9801 + 2025 + 3025 = 14851$.

Answer: 14851

Problem 9 Solution
In order to find the maximum a, we need to find the powers of 2 and 3 in the prime factorization of 100!. The power of 2 is

$$\left\lfloor \frac{100}{2} \right\rfloor + \left\lfloor \frac{100}{2^2} \right\rfloor + \left\lfloor \frac{100}{2^3} \right\rfloor + \left\lfloor \frac{100}{2^4} \right\rfloor + \left\lfloor \frac{100}{2^5} \right\rfloor + \left\lfloor \frac{100}{2^6} \right\rfloor$$

$$= 50 + 25 + 12 + 6 + 3 + 1$$

$$= 97,$$

and the power of 3 is

$$\left\lfloor \frac{100}{3} \right\rfloor + \left\lfloor \frac{100}{3^2} \right\rfloor + \left\lfloor \frac{100}{3^3} \right\rfloor + \left\lfloor \frac{100}{3^4} \right\rfloor$$

$$= 33 + 11 + 3 + 1$$

$$= 48.$$

Since

$$2^{97} \times 3^{48} = 12^{48} \times 2,$$

the maximum power of 12 is 48. Thus $a_{max} = 48$.

Answer: 48

Problem 10 Solution
Extend $\overline{DC}$ to point G so that $CG = CE$, as shown.

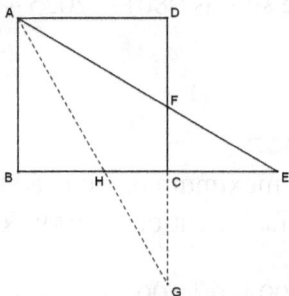

Answer: 2

Problem 11 Solution
Let x_1, x_2, x_3, and x_4 be the four roots of the equation, in increasing order, so they form an arithmetic progression. Clearly, if r is a root of the equation, so is $-r$, therefore

$$x_1 = -x_4, \quad x_2 = -x_3.$$

The equation can be transformed to

$$x^4 - 5x^2 + 4 - k = 0,$$

thus by Vieta's Formulas,

$$x_1x_2 + x_1x_3 + x_1x_4 + x_2x_3 + x_2x_4 + x_3x_4 = -5.$$

Since $x_1 = -x_3$ and $x_2 = -x_4$,

$$-x_3^2 - x_4^2 = -5,$$

which is

$$x_3^2 + x_4^2 = 5.$$

We know that x_1, x_2, x_3, x_4 is an arithmetic progression, so

$$x_3 - x_2 = x_4 - x_3,$$

which means

$$2x_3 = x_4 - x_3,$$

thus

$$x_4 = 3x_3.$$

So

$$x_3^2 + 9x_3^2 = 10x_3^2 = 5,$$

hence

$$x_3 = \frac{1}{\sqrt{2}}.$$

so

$$x_4 = \frac{3}{\sqrt{2}}, \quad x_1 = -\frac{3}{\sqrt{2}}, \quad x_2 = -\frac{1}{\sqrt{2}}.$$

Also by Vieta's Formulas,

$$x_1x_2x_3x_4 = 4 - k,$$

hence

$$k = 4 - \frac{9}{4} = \frac{7}{4} = 1.75.$$

Answer: 1.75

Problem 12 Solution

Connect $\overline{CE}$ and $\overline{AC}$, as shown.

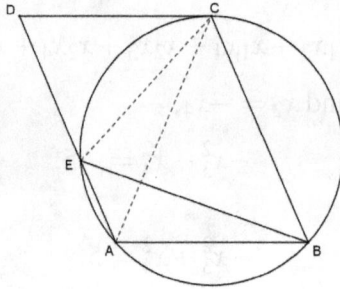

Answer: 3.2

Problem 13 Solution

Let a, b, c, and d denote the number of golf balls Phil puts into each of the 4 boxes. If we let e denote how many he has left over, then we want

$$a+b+c+d+e = 10 \text{ with } e \neq 20.$$

Using stars and bars there are

$$\binom{10+5-1}{10} = \binom{14}{10}$$
$$= \frac{14 \cdot 13 \cdot 12 \cdot 11}{4!}$$
$$= 7 \cdot 13 \cdot 11$$
$$= 1001$$

outcomes with $a+b+c+d+e = 10$. As there is one outcome with $e = 20$, there are $1001 - 1 = 1000$ total outcomes.

Answer: 1000

Problem 14 Solution

Recursively consider H_{n+1} drawn inside H_n as shown below.

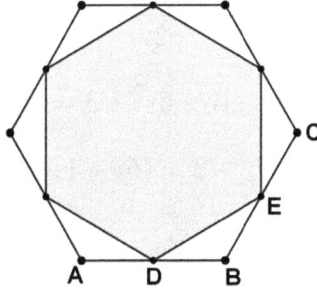

Here D and E are midpoints of $\overline{AB}$ and $\overline{BC}$ respectively. For easier calculations assume $AB = BC = 4x$ so $BD = BE = 2x$. Note that $\triangle DBE$ is isosceles with $\angle DBE = 120°$, so breaking $\triangle DBE$ into two 30-60-90 triangles we see

$$DE = 2 \cdot x\sqrt{3} = 2x\sqrt{3}.$$

Both H_{n+1} and H_n are regular, so as their ratio of sides is $2\sqrt{3} : 4$ the ratio of their areas is

$$(2\sqrt{3})^2 : 4^2 = 12 : 16 = 3 : 4.$$

Since H_0 has area 1, repeated applications of this give that H_n has area $\left(\dfrac{3}{4}\right)^n$. Calculating powers of this we have

$$\frac{3}{4}, \frac{9}{16}, \frac{27}{64}, \frac{81}{256}, \frac{243}{1024}$$

so H_5 is the first with area ≤ 0.25. Our answer is $n = 5$.

Answer: 5

Problem 15 Solution

Making the substitution $y = \dfrac{x+2}{x+1}$ we get

$$y - \frac{1}{2y} = \frac{17}{6}.$$

Clearing denominators gives $6y^2 - 3 = 17y$ so factoring we have

$$6y^2 - 17y - 3 = (6y+1)(y-3) = 0$$

and thus $y = -\dfrac{1}{6}$ or $y = 3$. Hence

$$\frac{x+2}{x+1} = -\frac{1}{6} \Rightarrow x = -\frac{13}{7}$$

or

$$\frac{x+2}{x+1} = 3 \Rightarrow x = -\frac{1}{2}.$$

The smallest of these solutions is $-\dfrac{13}{7}$ so $P + Q = -13 + 7 = -6$.

Answer: -6

Problem 16 Solution

Because Raul is dealt 7 physical cards, for probability purposes we consider the two jacks distinguishable. Therefore the total number of outcomes is $7!$.

We want the numbered cards to appear in order, so there is only one possible way for them to be ordered. Hence we divide the total number of outcomes by $5!$ to count the outcomes we want. Hence there are $\dfrac{7!}{5!} = 7 \cdot 6$ outcomes with the numbers in order. This gives a probability of

$$\frac{7 \cdot 6}{7!} = \frac{1}{5!} = \frac{1}{120}$$

so $Q - P = 120 - 1 = 119$.

Answer: 119

Problem 17 Solution

To find the two intersection points, we solve the equation

$$x^2 - 2ax + a^2 + 2a = 2x + 4,$$

which is

$$x^2 - (2a + 2)x + (a^2 + 2a - 4) = 0.$$

Using the Quadratic Formula,

$$x_1 = a + 1 + \sqrt{5}, \quad x_2 = a + 1 - \sqrt{5},$$

thus

$$y_1 = 2a + 6 + 2\sqrt{5}, \quad y_2 = 2a + 6 - 2\sqrt{5}.$$

Hence

$$AB^2 = (x_1 - x_2)^2 + (y_1 - y_2)^2 = (2\sqrt{5})^2 + (4\sqrt{5})^2 = 100,$$

therefore

$$AB = 10,$$

and it is the only possible value.

Answer: 10

Problem 18 Solution

Clear the denominators, the equation becomes

$$y(x + y) - x(x + y) = xy,$$

so

$$xy + y^2 - x^2 - xy - xy = 0,$$

which is

$$y^2 - xy - x^2 = 0.$$

Dividing by x^2,

$$\left(\frac{y}{x}\right)^2 - \left(\frac{y}{x}\right) - 1 = 0.$$

Since x and y are positive,

$$\frac{y}{x} = \frac{1+\sqrt{5}}{2}.$$

So we have

$$\frac{x}{y} = \frac{\sqrt{5}-1}{2}.$$

Let $t = \frac{y}{x}$, we have

$$t + \frac{1}{t} = \sqrt{5},$$

and

$$\left(t+\frac{1}{t}\right)^3 = t^3 + \frac{1}{t^3} + 3t + \frac{3}{t},$$

therefore

$$\begin{aligned}
\left(\frac{x}{y}\right)^3 + \left(\frac{y}{x}\right)^3 &= t^3 + \frac{1}{t^3} \\
&= \left(t+\frac{1}{t}\right)^3 - 3\left(t+\frac{1}{t}\right) \\
&= 5\sqrt{5} - 3\sqrt{5} \\
&= 2\sqrt{5}.
\end{aligned}$$

Therefore $P = 2$ and $Q = 5$, and $P + Q = 7$.

Answer: 7

Problem 19 Solution

Rationalizing a_n,

$$a_n = \frac{1}{\sqrt{n} + \sqrt{n+1}} = \sqrt{n+1} - \sqrt{n}.$$

Thus

$$S_n = \sqrt{2} - \sqrt{1} + \sqrt{3} - \sqrt{2} + \cdots + \sqrt{n+1} - \sqrt{n} = \sqrt{n+1} - 1.$$

Since

$$S_M = \sqrt{M+1} - 1 = 18,$$

we have

$$M + 1 = 19^2 = 361,$$

therefore $M = 360$.

Answer: 360

Problem 20 Solution

From the second equation, $(x+y)z = 17$. Since 17 is a prime, and x, y, z are all positive integers, $z = 1$, and $x + y = 17$. Thus the system of equations becomes

$$\begin{cases} xy + y &= 45, \\ x + y &= 17. \end{cases}$$

Hence,

$$xy - x = 45 - 17 = 28,$$

which means

$$(y - 1)x = 28.$$

So the possible values of x are 1, 2, 4, 7, 14, and 28, and the corresponding values of y are, respectively, 29, 15, 8, 5, 3, and 2. Considering $x + y = 17$, only two solutions exist: $(x,y) = (2, 15)$, and $(x,y) = (14, 3)$.

If $(x,y,z) = (2, 15, 1)$, $x^2 + y^2 + z^2 = 230$;

If $(x,y,z) = (14,3,1)$, $x^2 + y^2 + z^2 = 206$.

Therefore the answer is 230.

Answer: 230

2.5   ZIML February 2019 Junior Varsity

Below are the solutions from the Junior Varsity ZIML Competition held in February 2019.
The problems from the contest are available on p.49.

Problem 1 Solution
Since 40% of the people surveyed do not regularly drink soda, then 60% regularly drink soda. Since 70% regularly drink both,

$$50\% + 60\% - 70\% = 40\%$$

regularly drink coffee and soda. Therefore

$$70\% - 40\% = 30\%$$

drink either coffee or soda (but not both). Hence $K = 30$.

Answer: 30

Problem 2 Solution
Factoring we have

$$75 = 3 \cdot 5^2, 130 = 2 \cdot 5 \cdot 13, \text{ and } 140 = 2^2 \cdot 5 \cdot 7.$$

Therefore

$$\text{lcm}(75, 130, 140) = 2^2 \cdot 3 \cdot 5^2 \cdot 7 \cdot 13 = 27300.$$

Calculating $2020000 \div 27300 = 73$ with remainder 27100. Therefore the smallest number divisible by 27300 is

$$2020000 - 27100 + 27300 = 2020200,$$

which is our answer.

Answer: 2020200

Problem 3 Solution

If Luke lines up the 7 basketballs, he creates $7 + 1 = 8$ spaces he can put the trophies. Hence there are

$$8 \cdot 7 \cdot 6 \cdot 5 = 1680$$

arrangements.

Answer: 1680

Problem 4 Solution

Using Vieta's theorem we know the sum of the roots is $-(-4) = 4$ and the product of the roots is $-(-6) = 6$. The only way to write 6 as the product of three positive integers is

$$1 \cdot 2 \cdot 3 \text{ or } 1 \cdot 1 \cdot 6.$$

It is impossible for all numbers to be positive and have a sum of 4. The other possibility is that two are negative, and we see that -1, -1, and 6 works, and hence are the roots. Using Vieta's theorem again we have

$$C = -1 \cdot -1 + -1 \cdot 6 + -1 \cdot 6 = -11.$$

Therefore $C = -11$ is our answer.

Answer: -11

Problem 5 Solution

We first calculate the area of the full octagon. Adding four 45-45-90 triangles to the octagon, we get a square as shown below.

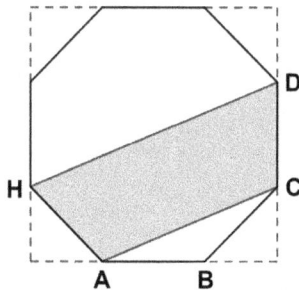

Since the octagon has side lengths of 1, each of these triangles has sides of $\frac{\sqrt{2}}{2}$, $\frac{\sqrt{2}}{2}$, 1. Hence the area of the octagon is

$$(1+\sqrt{2})^2 - 4 \cdot \frac{1}{2} \cdot \left(\frac{\sqrt{2}}{2}\right)^2$$

$$= 3 + 2\sqrt{2} - 1$$

$$= 2\sqrt{2} + 2$$

Therefore, $[ABCDH] = \sqrt{2} + 1$ as it is half the octagon. Further, we know $\triangle ABC$ has a base of 1 and height $\frac{\sqrt{2}}{2}$ so $[ABC] = \frac{1}{2} \cdot 1 \cdot \frac{\sqrt{2}}{2} = \frac{\sqrt{2}}{4}$. Combining we have

$$= [ABCDH] - [ABC]$$

$$= \sqrt{2} + 1 - \frac{\sqrt{2}}{4}$$

$$= 1 + \frac{3}{4}\sqrt{2}$$

and therefore $A + B + C = 1 + 0.75 + 2 = 3.75$.

Answer: 3.75

Problem 6 Solution

Factoring, $180 = 2^2 \cdot 3^2 \cdot 5$ so $180^{20} = 2^{40} \cdot 3^{40} \cdot 5^{20}$. Therefore 180^{20} has

$$(40+1)(40+1)(20+1) = 41 \cdot 41 \cdot 21$$

factors.

However, we need to subtract off the factors that are divisible by 10. Any factor that is divisible by 10 can be written as $10 \cdot F$ where F is a factor of $2^{39} \cdot 3^{40} \cdot 5^{19}$. Hence there are

$$(39+1)(40+1)(19+1) = 41 \cdot 40 \cdot 20$$

factors that are a multiple of 10.

This gives us

$$
\begin{aligned}
& 41 \cdot 41 \cdot 21 - 41 \cdot 40 \cdot 20 \\
&= 41(41 \cdot 21 - 40 \cdot 20) \\
&= 41(861 - 800) \\
&= 41 \cdot 61 \\
&= 2501
\end{aligned}
$$

factors of 180^{20} that are not divisible by 10.

Answer: 2501

Problem 7 Solution

Making the substitution $z = 2^x$ we have (as $4^x = z^2$): $z^2 - 3z + 4$. This is a quadratic with vertex when

$$z = -\frac{-3}{2} = \frac{3}{2}.$$

Therefore the minimum of $z^2 - 3z + 4$ is

$$\left(\frac{3}{2}\right)^2 - 3 \cdot \frac{3}{2} + 4$$
$$= \frac{9}{4} - \frac{9}{2} + 4$$
$$= \frac{9 - 18 + 16}{4}$$
$$= \frac{7}{4}.$$

Since it is possible for $2^x = \frac{3}{2}$, the minimum of $f(x)$ is also $\frac{7}{4} = 1.75$.

Answer: 1.75

Problem 8 Solution

To maximize area, we want the side lengths to be as close to each other as possible. In this case, this means they are consecutive integers, so they must be 9, 10, and 11.

Using Heron's formula, the area is

$$\sqrt{15(15-9)(15-10)(15-11)}$$
$$= \sqrt{15 \cdot 6 \cdot 5 \cdot 4}$$
$$= \sqrt{1800}$$

and $K = \sqrt{1800}$.

Since $40^2 = 1600$, we start checking squares above 40. We find

$$42^2 < 1800 < 43^2$$

so $\lfloor K \rfloor = 42$.

Answer: 42

Problem 9 Solution
Using Vieta's theorem we know

$$r+s = -\frac{16}{4} = -4 \text{ and } r\cdot s = \frac{9}{4}.$$

Noting that
$$(r+s)^3 = r^3 + 3r^2s + 3rs^2 + s^3$$
$$= r^3 + s^3 + 3rs(r+s),$$

we have
$$r^3 + s^3 = (r+s)^3 - 3rs(r+s)$$
$$= (-4)^3 - 3\cdot\frac{9}{4}\cdot -4$$
$$= -64 + 27$$
$$= -37.$$

Therefore we have $P+Q = -37+1 = -36$.

Answer: -36

Problem 10 Solution
Using the alternating sum

$$N \equiv 9-1+0-2+9-1+0-2\cdots$$
$$\equiv 9\cdot(9-1+0-2)$$
$$\equiv 9\cdot 6$$
$$\equiv 54$$
$$\equiv -1 \quad (\text{mod } 11).$$

Therefore

$$N \equiv (-1)^{2019} \equiv -1 \equiv 10 \quad (\text{mod } 11)$$

so the remainder when divided by 11 is 10.

Answer: 10

Problem 11 Solution

If C is on the perpendicular bisector of $\overline{AB}$, then C is equidistant from A and B. Hence B must be on the circle containing A with center C. The radius of this circle is

$$\sqrt{(-4-(-2))^2+(1-4)^2} = \sqrt{13}$$

so it has equation

$$(x+2)^2 + (y-4)^2 = 13.$$

Therefore we want integers r, s with $r > 0$ such that

$$(r+2)^2 + (s-4)^2 = 13.$$

The only integers whose squares sum to 13 are 2 and 3. As $r > 0$ we must have $r+2 = 3$ and $r = 1$. Hence $|s-4| = 2$ so either $s = 6$ or $s = 2$. The smallest possible value of s is therefore 2.

Answer: 2

Problem 12 Solution

The distance between $(0, a)$ and $(b, 0)$ is $\sqrt{a^2 + b^2}$, so we are looking for the probability of picking two numbers $0 \le a, b \le 1$ with $\sqrt{a^2 + b^2} \le \dfrac{1}{2}$.

Consider a unit square. A point on this square represents the possible pairs (a, b) that can be chosen. Note the region of the square that contains the points where $a^2 + b^2 \le \dfrac{1}{2}$ is a quarter circle with radius $\dfrac{1}{2}$, which has area $\dfrac{\pi}{16}$. Since the square has area 1, the probability of choosing points with distance at most $\dfrac{1}{2}$ is $\dfrac{\pi}{16}$. Therefore $R \cdot S = 1 \cdot 16 = 16$.

Answer: 16

Problem 13 Solution

Construct a right triangle ADB where $AD = 5$ and $DB = 15$, with hypotenuse $AB = \sqrt{5^2 + 15^2} = 5\sqrt{10}$. Let E be the point on $\overline{AD}$ such that $AE = 3$, and F be the point on $\overline{DB}$ such that $DF = 5$. Let C be the point such that $DFCE$ is a rectangle, as shown.

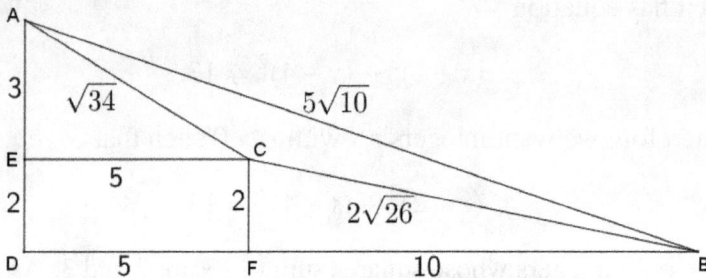

Thus, using $[ABC]$ to denote the area of $\triangle ABC$ (similarly for other regions),

$$
\begin{aligned}
[ABC] &= [ABD] - [ACE] - [CBF] - [DECF] \\
&= \frac{5 \times 15}{2} - \frac{3 \times 5}{2} - \frac{2 \times 10}{2} - 2 \times 5 \\
&= 10.
\end{aligned}
$$

Answer: 10

Problem 14 Solution

Using the binomial theorem we can calculate the coefficient of x^4 in $(x^2 + R)^4$ and in $(x + 4)^6$ separately. In $(x^2 + R)^4$ the term with x^4 is

$$
\binom{4}{2} \cdot (x^2)^2 \cdot R^2 = 6R^2 x^4.
$$

Similarly in $(x+4)^6$ the term with x^4 is

$$\binom{6}{2} \cdot x^4 \cdot 4^2 = 240x^4.$$

Since the combined coefficient must be 0 we have

$$6R^2 = 240 \Rightarrow R = \pm\sqrt{40}.$$

Thus

$$R^4 - 4^6 = 1600 - 4096 = -2496$$

is the constant term.

Answer: -2496

Problem 15 Solution

The remainders of $3^0, 3^1, 3^2, \ldots$ when divided by 40 are

$$1, 3, 9, 27, 1, 3, 9, 27, \ldots$$

and this pattern repeats every 4 terms. Since Christopher adds up these remainders, his answer is

$$\begin{aligned} C &= 1 + 3 + 9 + 27 + \cdots + 1 \\ &= 5 \cdot (1 + 3 + 9 + 27) + 1 \\ &= 5 \cdot 40 + 1 \\ &= 201 \end{aligned}$$

Note Christopher's method would lead to the correct answer if he just took the remainder of this when divided by 40. Therefore Michael's answer $M = 1$ so $C - M = 200$.

Answer: 200

Problem 16 Solution

Note $8 = 2^3$, $27 = 3^3$, $125 = 5^3$ and $1331 = 11^3$ so the prime factorization is

$$35937000 = 2^3 \cdot 3^3 \cdot 5^3 \cdot 11^3.$$

The three powers of 2 must be distributed among the 4 integers, which can be done using stars and bars. Identical reasoning for the other primes gives that there are

$$\binom{3+4-1}{3}^4 = \binom{6}{3}^4 = 20^4 = 160000$$

ways to write the number as the product of 4 positive integers.

Answer: 160000

Problem 17 Solution

There is a $\dfrac{1}{2}$ chance that she will get heads and a $\dfrac{1}{2}$ chance that she will get tails.

If she gets heads, she rolls once and there is a $\dfrac{1}{3}$ chance that she gets a number more than 4.

If she gets tails, she rolls three times and the only way to get a sum ≤ 4 is if the sum is 3 ($1+1+1$, so 1 way) or 4 ($2+1+1$, so 3 ways after reordering). Hence there is a

$$1 - \frac{4}{6^3} = 1 - \frac{1}{54} = \frac{53}{54}$$

chance if she gets tails.

Therefore, the probability that she gets a sum larger than 2 is

$$\frac{1}{2} \times \frac{1}{3} + \frac{1}{2} \times \frac{53}{54}$$
$$= \frac{1}{6} + \frac{53}{108}$$
$$= \frac{18+53}{108}$$
$$= \frac{71}{108}.$$

Thus, $Q - P = 108 - 71 = 37$.

Answer: 37

Problem 18 Solution

$\overline{AB}$ is a diameter, so it divides the circle into two semicircles. Hence $\angle ACB = 90°$. Further, since $\overparen{BC} = 60°$, we have $\angle BAC = 60° \div 2 = 30°$ so $\triangle ABC$ is a 30-60-90 triangle.

Using $\overline{BC} \parallel \overline{DE}$, $\triangle AFG$ is also a 30-60-90 triangle, and as $\triangle CDG$ is isosceles, we know it is a 45-45-90 triangle. Further, $\overline{BC} \parallel \overline{DE}$ also implies that $\overparen{BE} = \overparen{CD}$, so $\angle BED = \angle CDE = 45°$.

Putting all the angles together:

$$\angle ABE = 180° - \angle BED - \angle BFE$$
$$= 180° - 45° - 60°$$
$$= 75°.$$

Hence $\overparen{AE} = 2 \cdot 75° = 150°$ so our answer is 150.

Answer: 150

Problem 19 Solution

To save a little time, consider the prime factorization of 21504 which is

$$21504 = 2^{10} \cdot 3 \cdot 7.$$

Hence we see $a_2 = 2^9 \cdot 21$, $a_3 = 2^8 \cdot 21$, etc., until $a_{11} = 21$. Then

$$a_{12} = 21 \cdot 3 + 1 = 64 = 2^6.$$

Repeating the strategy from above we see $a_{18} = 1$ so $M = 18$.

Answer: 18

Problem 20 Solution

We consider cases based on $|Kx + 3|$.

If $Kx + 3 > 0$ we have

$$Kx + 3 = x^2 + Kx + 1 \Rightarrow x^2 = 2$$

so $x = \pm\sqrt{2}$. For $x = \sqrt{2}$ to be a solution, we need

$$K\sqrt{2} + 3 > 0 \Rightarrow K > -\frac{3}{\sqrt{2}}.$$

Identical reasoning gives $x = -\sqrt{2}$ is a solution when $K < \dfrac{3}{\sqrt{2}}$.

As $\sqrt{2} \approx 1.41$, $x = \sqrt{2}$ is a solution for integers $K \geq -2$ and $x = -\sqrt{2}$ is a solution for integers $K \leq 2$.

If $Kx + 3 < 0$ we have

$$-Kx - 3 = x^2 + Kx + 1 \Rightarrow x^2 + 2Kx + 4 = 0.$$

The discriminant is

$$(2K)^2 - 4 \cdot 4 = 4(K^2 - 4).$$

Hence this has 2 solutions when $|K| > 2$ and 1 solution when $|K| = 2$.

Combining with the other case, we see there are more than 2 solutions only when $|K| = 2$. Hence the product of all possible K is $-2 \cdot 2 = -4$.

(Double checking, when $K = -2$ we have solutions $x = 2, \pm\sqrt{2}$ and when $K = 2$ we have solutions $x = -2, \pm\sqrt{2}$.)

Answer: -4

2.6 ZIML March 2019 Junior Varsity

Below are the solutions from the Junior Varsity ZIML Competition held in March 2019.

The problems from the contest are available on p.57.

Problem 1 Solution
To find the prime factorization of 234432 we use divisibility rules.

The sum of the digits is

$$2+3+4+4+3+2 = 18$$

so the number is divisible by 9. The alternating sum of the digits is

$$2-3+4-4+3-2 = 0$$

so the number is also divisible by 11. Hence 234432 is divisible by 99 with $234432 \div 99 = 2368$. Repeated division by 2 gives $2368 = 2^6 \cdot 37$. Thus

$$234432 = 2^6 \cdot 3^2 \cdot 11 \cdot 37$$

which has $(6+1)(2+1)(1+1)(1+1) = 84$ factors.

Answer: 84

Problem 2 Solution
Factoring,

$$\frac{a^4 - b^4}{a-b} = \frac{(a-b)(a+b)(a^2+b^2)}{(a-b)} = (a+b)(a^2+b^2).$$

Using Vieta's theorem $a+b = -\dfrac{6}{1} = -6$ and $ab = \dfrac{2}{1} = 2$. Thus

$$\begin{aligned}
(a+b)(a^2+b^2) &= (a+b)[(a+b)^2 - 2ab] \\
&= -6 \cdot [(-6)^2 - 2 \cdot 2] \\
&= -6 \cdot 32 \\
&= -192
\end{aligned}$$

which gives the answer.

Answer: -192

Problem 3 Solution
From 1 to 9 there are 5 odd numbers and 4 even numbers. The first card (which is removed) is either odd or even, so consider two cases.

Both cards being odd has probability

$$\frac{5}{9} \cdot \frac{4}{8} = \frac{5}{18}.$$

The first card even and second card odd has probability

$$\frac{4}{9} \cdot \frac{5}{8} = \frac{5}{18}.$$

Combined this is probability $\frac{5}{9}$ so $Q - P = 9 - 5 = 4$.

Answer: 4

Problem 4 Solution
For $\overline{abc}$ to be a three-digit number we need $a \neq 0$. We have

$$\overline{ab} + c \equiv -a + b + c \pmod{11}.$$

Note this is the alternating sum of $\overline{bac}$.

Ignoring now that a cannot be zero, the above shows that counting $\overline{ab} + c$ that are multiples of 11 is equivalent to counting $\overline{bac}$ that are multiples of 11. $999 \div 11 = 90$ with remainder 9, so there are 90 multiples of 11 from 1 to 999.

If $a = 0$, $\overline{bac}$ is a multiple of 11 if $b + c = 11$. Thus

$$(b,c) = (2,9),(3,8),\ldots(9,2)$$

so there are 8 multiples of 11 with $a = 0$ that we need to subtract off. This gives an answer of $90 - 8 = 82$.

Answer: 82

Problem 5 Solution

Since AD is the angle bisector of $\angle CAB$, the Angle Bisector Theorem says that $\dfrac{AC}{CD} = \dfrac{AB}{BD}$. Thus, $\dfrac{BD}{CD} = \dfrac{3}{5}$, so the segments BD and CD are in ratio $3 : 5$.

Since $BD + DC = 24$, and $3 : 5 = 9 : 15$, we have $BD = 9$ and $DC = 15$.

Answer: 9

Problem 6 Solution

The lines $y = x + 1$ and $y = x - 2$ are parallel. Thus a point in the plane is closer to Irene's line if it is below the line $y = x - 0.5$, which is halfway between the two lines. Hence (a, b) is closer to Irene's line if it is in the shaded region shown below

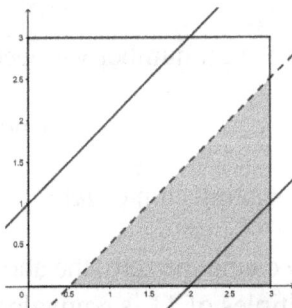

The line $y = x - 0.5$ intersects the square at $(0.5, 0)$ and $(3, 2.5)$, so the area of the square closer to Irene's line is

$$\frac{1}{2} \cdot 2.5^2 = \frac{1}{2}\left(\frac{5}{2}\right)^2 = \frac{25}{8}.$$

The full square has area 9, giving probability

$$\frac{25}{8} \div 9 = \frac{25}{72}$$

with $P + Q = 25 + 72 = 97$.

Answer: 97

Problem 7 Solution

Inside the square root we have $\sqrt{9 - x^2}$ which is defined only for x such that $-3 \leq x \leq 3$. For the outer square root to be defined we need $4x - \sqrt{9 - x^2} \geq 0$. Since $\sqrt{9 - x^2}$ is always non-negative, $x \geq 0$. Hence we must have

$$4x \geq \sqrt{9 - x^2}$$
$$16x^2 \geq 9 - x^2$$
$$17x^2 \geq 9$$
$$x^2 \geq \frac{9}{17}$$
$$x \geq \frac{3}{\sqrt{17}}.$$

Hence $L = \dfrac{3}{\sqrt{17}} = \dfrac{3\sqrt{17}}{17}$ so $R + S + T = 3 + 17 + 17 = 37$.

Answer: 37

Problem 8 Solution

Since chords $\overline{BD}$ and $\overline{CE}$ bisect $\overline{AB}$ and $\overline{AC}$ respectively, $\overline{BD}$ and $\overline{CE}$ are each extensions of medians in $\triangle ABC$, so F (their intersection) is the centroid of the triangle.

As $\overline{AB}$ is the hypotenuse of the right triangle, it is a diameter, so

$\overline{CE}$ is a diameter as well. Therefore

$$CF = \frac{2}{3} \cdot 10 = \frac{20}{3}$$
$$\text{and } EF = 20 - CF = \frac{40}{3}.$$

Using power of a point we therefore have that

$$BF \cdot DF = CF \cdot EF$$
$$8 \cdot DF = \frac{20}{3} \cdot \frac{40}{3}$$
$$DF = \frac{100}{9}$$

so $P - Q = 100 - 9 = 91$.

Answer: 91

Problem 9 Solution
Rewriting we have

$$9x^2 + 6x + 1 = 2|3x+1|.$$

As $9x^2 + 6x + 1 = (3x+1)^2$ making the substitution $u = |3x+1|$ we have
$$u^2 = 2u$$

so either $u = 0$ or $u = 2$.

If $u = 0$ we have $|3x+1| = 0$ so $3x+1 = 0$ and $x = -\frac{1}{3}$.

If $u = 2$ we have $|3x+1| = 2$ so $3x+1 = 2$ and $x = \frac{1}{3}$ or $3x+1 = -2$ and $x = -1$.

Double checking, all three are roots, with

$$\left(-\frac{1}{3}\right)^2 + \left(\frac{1}{3}\right)^2 + 1^2$$
$$= \frac{1}{9} + \frac{1}{9} + 1$$
$$= \frac{11}{9}.$$

Therefore $P + Q = 11 + 9 = 20$.

Answer: 20

Problem 10 Solution

Let I be 6 minus the first roll (which decides how many dice Ben rolls). Using the sample rows from the question, we would then have

I	$1s$	$2s$	$3s$	$4s$	$5s$	$6s$
4	2	0	0	0	0	0
0	0	1	3	1	1	0

Note in this expanded table, each row sums to 6, as if x is the first roll, we roll the die x times (so we get x total 1s, 2s, up to 6s) and $I = 6 - x$.

Furthermore, I is already determined by the other rolls, so there is no new information in the expanded table. Using stars and bars (we have 7 columns and the sum is 6) there are

$$\binom{6+7-1}{6} = \binom{12}{6} = 924$$

possible rows. However, one of these outcomes $(6, 0, \ldots, 0)$ is impossible. Hence the answer is $924 - 1 = 923$.

Answer: 923

Problem 11 Solution

To find the last digit work modulo 10. This gives

$$x_{n+1} = 9 \cdot x_n + 1 \equiv -x_n + 1 \quad (\text{mod } 10).$$

Therefore

$$x_{n+2} \equiv -x_{n+1} + 1 \quad (\text{mod } 10)$$
$$\equiv -(-x_n + 1) + 1 \quad (\text{mod } 10)$$
$$\equiv x_n \quad (\text{mod } 10).$$

Hence the last digits of the term alternate. As $x_0 = I$ and $x_{2019} = 4$, they alternate between I and 4. Thus

$$4 \equiv -I + 1 \quad (\text{mod } 10)$$

so $I \equiv -3 \equiv 7 \pmod{10}$. Therefore, $I = 7$.

Answer: 7

Problem 12 Solution

Since $x = 0$ is not a solution, we can divide by x^2 to get

$$x^2 - 3x + 4 - 3\frac{1}{x} + \frac{1}{x^2} = 0.$$

Noticing the symmetry in the coefficients we make the substitution $u = x + \frac{1}{x}$, so $u^2 = x^2 + 2 + \frac{1}{x^2}$. Hence our equation becomes

$$u^2 - 3u + 2 = (u - 2)(u - 1) = 0.$$

Therefore $u = 2$ or $u = 1$.

If $u = 2$ we have

$$x + \frac{1}{x} = 2 \Rightarrow x^2 - 2x + 1 = (x - 1)^2 = 0$$

so $x = 1$ is a solution. If $u = 1$ we have

$$x + \frac{1}{x} = 1 \Rightarrow x^2 - x + 1 = 0$$

which has no real roots.

Thus $x = 1$ is the only real root, so our answer is 1.

Answer: 1

Problem 13 Solution

If a number has last digit 0 written in base B, then the number is divisible by B. Hence N is divisible by 2, 3, and 5, so N is divisible by $2 \cdot 3 \cdot 5 = 30$.

The last two digits of N are 22 in base 8, so modulo $8^2 = 64$,

$$N \equiv 2 \cdot 8 + 2 = 18 \pmod{64}.$$

Working mod 30, $18 \equiv -12 \pmod{30}$ and $64 \equiv 4 \pmod{30}$. As $-12 + 4 \cdot 3 = 0$,

$$N = 18 + 64 \cdot 3 = 210$$

is divisible by 30 and has remainder 18 when divided by 64. This gives $N = 210$ as the answer.

Answer: 210

Problem 14 Solution

Consider the diagram below

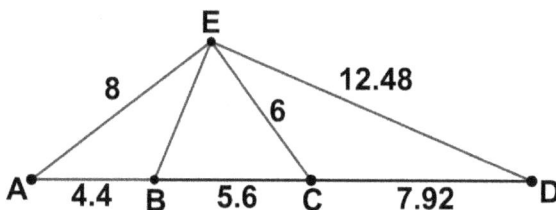

$AC = 4.4 + 5.6 = 10$, so looking at $\triangle AEC$ it is a 6-8-10 triangle, hence $\angle AEC = 90°$. As $\angle AEB = \angle DEC$, this means $\angle BED = 90°$ as well.

For right triangle $\triangle BED$, $DE = 12.48$ and $BD = 5.6 + 7.92 = 13.52$. Therefore

$$\frac{BD}{DE} = \frac{13.52}{12.48} = \frac{13\frac{13}{25}}{12\frac{12}{25}} = \frac{13}{12}$$

after dividing by $1\frac{1}{25} = 1.04$. Thus $\triangle BED$ is similar to a 5-12-13 triangle and $BE = 5 \cdot 1.04 = 5.2$.

Hence the perimeter of $\triangle BCE$ is $5.2 + 5.6 + 6 = 16.8$.

Answer: 16.8

Problem 15 Solution

By the Binomial Theorem,

$$\left(x^3 + \frac{1}{x\sqrt{x}}\right)^n$$
$$= (x^3 + x^{-1.5})^n$$
$$= \binom{n}{n}x^{3n} + \binom{n}{n-1}x^{3(n-1)-1.5} + \cdots + \binom{n}{0}x^{-1.5n}$$

where the kth term is

$$\binom{n}{n-k}x^{3(n-k)-1.5k}.$$

Therefore the constant term occurs when $3(n-k) - 1.5k = 0$ so $k = \frac{2}{3}n$.

Hence n must be a multiple of 3, so set $n = 3l$ for some l. Thus $k = 2l$ and the constant term is

$$\binom{3l}{l} = 84 = \binom{9}{3}.$$

Thus $3l = n = 9$.

Answer: 9

Problem 16 Solution

Write $p(x) = x^3 + bx^2 + cx + d$. Recalling that the remainder when $p(x)$ is divided by $(x - k)$ is equal to $p(k)$ (the Remainder Theorem) we get

$$1 + b + c + d = 4$$
$$d = 6$$
$$-1 + b - c + d = 0.$$

Adding the first and third equations we get $2b + 2d = 4$ so since $d = 6$ we have $b = -4$. Hence we solve for $c = 1$ and

$$p(x) = x^3 - 4x^2 + x + 6 = 0.$$

We know $x = -1$ is one root and as

$$\frac{p(x)}{x+1} = x^2 - 5x + 6 = (x-2)(x-3)$$

we have $x = 2$ and $x = 3$ as the other two roots. Therefore the largest root is 3.

Answer: 3

Problem 17 Solution

The base area is $18\sqrt{3}$, hence using the area formula for a regular hexagon

$$\frac{3s^2\sqrt{3}}{2} = 18\sqrt{3}$$

implying the side length $s = \sqrt{12} = 2\sqrt{3}$.

The lateral surface area is $48\sqrt{3} - 18\sqrt{3} = 30\sqrt{3}$, so each the isosceles triangle faces of the pyramid have area $30\sqrt{3} \div 6 = 5\sqrt{3}$.

As the base is $2\sqrt{3}$ the height is

$$\frac{2 \cdot 5\sqrt{3}}{2\sqrt{3}} = 5.$$

This is the hypotenuse of the right triangle shown in the diagram below (including the midpoint of a side, the center of the hexagon, and the apex of the pyramid):

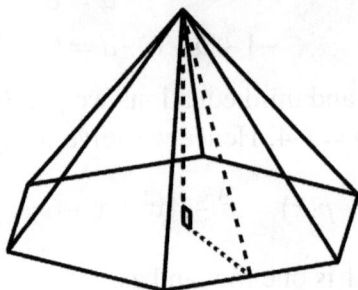

One of the sides is the height of the pyramid. The other side is the length of the altitude of a equilateral triangle with side length $2\sqrt{3}$ so 3. Thus this is a 3-4-5 triangle and the height of the pyramid is 4. This gives a volume of

$$\frac{1}{3} \cdot 18\sqrt{3} \cdot 4 = 24\sqrt{3}$$

with $K = 24$.

Answer: 24

Problem 18 Solution
Caution: $2019 = 3 \cdot 673$ so we cannot directly apply Fermat's Little Theorem. Instead we calculate 2017^{2018} (mod 3) and 2017^{2018} (mod 673) and by the Chinese Remainder Theorem we can find the remainder when 2017^{2018} is divided by 2019.

First note
$$2017^{2018} \equiv 1^{2018} \equiv 1 \pmod 3.$$

Using Fermat's Little Theorem, 2017^{672} (mod 673). Hence

$$2017^{2018} \equiv 2017^2 \equiv 2^2 \equiv 4 \pmod{673}.$$

Noting that 4 is equivalent to 1 (mod 3) and 4 (mod 673), we see the remainder when 2017^{2018} is divided by 2019 is 4.

Answer: 4

Problem 19 Solution
Check several cases based on the number of digits to find a pattern.

1 digit: there is only 1 "M" number: 1.

2 digits: there are 2 "M" numbers: 12, 13.

3 digits: there are 3 "M" numbers: 121, 131, 132.

4 digits: there are 5 "M" numbers: 1212, 1213, 1312, 1313, 1323.

5 digits: there are 8 "M" numbers: 12121, 12131, 12132, 13121, 13131, 13132, 13231, 13232.

It is clear that the numbers form a Fibonacci sequence: 1, 2, 3, 5, 8, ... (Note: this fact can be proven by induction) The 12th term in this sequence is 233.

Answer: 233

Problem 20 Solution

Clearly two of the sides of the triangle have length 3 and 4. Let c be the length of the third side. Then Heron's formula says the triangle has area

$$A = \sqrt{\frac{3+4+c}{2} \cdot \frac{3+4-c}{2} \cdot \frac{3-4+c}{2} \cdot \frac{-3+4+c}{2}}$$
$$= \sqrt{\frac{7+c}{2} \cdot \frac{7-c}{2} \cdot \frac{-1+c}{2} \cdot \frac{1+c}{2}}$$
$$= \frac{\sqrt{(49-c^2)(c^2-1)}}{4}$$
$$= \frac{\sqrt{c^4+50c^2-49}}{4},$$

so $c^4 + 50c^2 - 49 = 455$.

Let $u = c^2$. Then we can rewrite the equation as

$$u^2 + 50u - 504 = (u-36)(u-14) = 0,$$

so $u = 36$ or $u = 14$. Therefore $c^2 = 36$ or $c^2 = 14$. Since c (the third triangle side) must be a positive integer, $c = 6$ so the area of the third square has area 36.

Answer: 36

2.7 ZIML April 2019 Junior Varsity

Below are the solutions from the Junior Varsity ZIML Competition held in April 2019.
The problems from the contest are available on p.63.

Problem 1 Solution

Since each person gets the same color as the person seated directly across from them, it is enough to determine how 4 white and 4 black cards are given to one half of the table. This can be done in $\binom{8}{4} = 70$ ways, which is our answer.

Answer: 70

Problem 2 Solution

Assume $a > 0$ is the leading coefficient of the first parabola, so it has equation

$$y + 3 = a(x+1)^2$$
$$\Rightarrow y = ax^2 + 2ax + a - 3$$

The second parabola hence has leading coefficient $-a$ with equation

$$y - 1 = -a(x-2)^2$$
$$\Rightarrow y = -ax^2 + 4ax - 4a + 1$$

Subtracting the two equations gives

$$2ax^2 - 2ax + 5a - 4 = 0.$$

We want this equation to have exactly one solution, so setting the discriminant equal to 0 gives

$$(-2a)^2 - 4(2a)(5a - 4) = 0$$
$$4a^2 - 40a^2 + 32a = 0$$
$$-36a^2 + 32a = 0$$
$$-4a(9a - 8) = 0$$

so $a = \dfrac{8}{9}$ (as $a = 0$ is not possible). Thus $P + Q = 8 + 9 = 17$.

Answer: 17

Problem 3 Solution

Only the perfect squares 1^2, 2^2, up to 10^2 have an odd number of factors. We group by their prime factorizations to count their number of factors.

1 has 1 factor and all the squares of primes (2^2, 3^2, 5^2, and 7^2) have $2 + 1 = 3$ factors.

$4^2 = 2^4$ and $9^2 = 3^4$ both have $4 + 1 = 5$ factors. $8^2 = 2^6$ has $6 + 1 = 7$ factors. Finally $6^2 = 2^2 \cdot 3^2$ and $10^2 = 2^2 \cdot 5^2$ have $(2 + 1)^2 = 9$ factors.

Hence adding up the numbers Terry gets

$$1 + 4 \cdot 3 + 2 \cdot 5 + 7 + 2 \cdot 9 = 48.$$

Answer: 48

Problem 4 Solution

To avoid factoring a quartic polynomial, note

$$(x + 4)(x - 5) = x^2 - x - 20$$
$$\text{and } (x + 1)(x - 2) = x^2 - x - 2$$

which differ only in their constant term. Making the substitution $u = x^2 - x - 2$ we get

$$(u - 18) + 5 = \frac{14}{u}$$
$$u^2 - 13u - 14 = 0$$
$$(u - 14)(u + 1) = 0$$

so $u = 14$ or $u = -1$. Hence

$$x^2 - x - 16 = 0 \text{ or } x^2 - x - 1 = 0.$$

The quadratic formula gives us respective solutions of

$$\frac{1 \pm \sqrt{65}}{2} \text{ and } \frac{1 \pm \sqrt{5}}{2},$$

and therefore $B = 65$ or $B = 5$, with sum $5 + 65 = 70$.

Answer: 70

Problem 5 Solution

Let O denote the center of the circle. Then the area of the region is made up of sectors $\overgroup{AC}$ and $\overgroup{BD}$ and triangles $\triangle AOD$ and $\triangle BOC$. Consider the diagram below

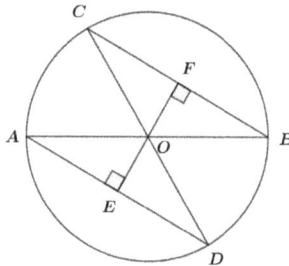

We know the sectors each measure $60°$ so their combined area is

$$2 \cdot \frac{60}{360} \cdot \pi 6^2 = \frac{36\pi}{3} = 12\pi.$$

Since $\angle AOC = 60°$, $\triangle AOD \cong \triangle BOC$ with angles of $120°$, $30°$, and $30°$. Thus $\overline{EF}$ divides these triangles into four congruent 30-60-90 triangles, each with hypotenuse 6 and hence legs of 3 and $3\sqrt{3}$. Therefore the combined area of $\triangle AOD$ and $\triangle BOC$ is

$$4 \cdot \frac{1}{2} \cdot 3 \cdot 3\sqrt{3} = 18\sqrt{3}.$$

Therefore the region has area

$$12\pi + 18\sqrt{3}$$

and $R + S + T = 12 + 18 + 3 = 33$.

Answer: 33

Problem 6 Solution

$x = \dfrac{1}{2}(-3 + \sqrt{13})$ so x and $\dfrac{1}{2}(-3 - \sqrt{13})$ are both roots of the same quadratic equation. Using Vieta's theorem, the linear term is opposite their sum

$$-\left(-\frac{3}{2} - \frac{3}{2}\right) = 3$$

and the constant term is their product

$$\frac{1}{4}(9 - 13) = -1.$$

Therefore $x^2 + 3x - 1 = 0$, so $x^2 + 3x = 1$. Grouping and rewriting,

$$\begin{aligned}
&x(x+1)(x+2)(x+3) \\
&= x(x+3) \cdot (x+1)(x+2) \\
&= (x^2 + 3x)(x^2 + 3x + 2) \\
&= 1 \cdot (1 + 2)
\end{aligned}$$

so the answer is 3.

Answer: 3

Problem 7 Solution

$2000 = 2^4 \cdot 5^3$ so 2000 has $(4+1)(3+1) = 20$ factors. Hence there are $20^2 = 400$ total ways to choose A and B.

Clearly if $A = 1$ or $B = 1$ (or both) then $\gcd(A, B) = 1$. This gives $20 + 20 - 1 = 39$ choices for A and B.

Else assume $A > 1$ and $B > 1$. Then if $\gcd(A, B) = 1$ we have $A = 2^k$ for $k = 1, 2, 3, 4$ and $B = 5^j$ for $j = 1, 2, 3$ or vice-versa. Therefore there are $4 \cdot 3 \cdot 2 = 24$ additional choices for A and B.

Hence the probability is

$$\frac{39 + 24}{400} = \frac{63}{400}$$

and $Q - P = 400 - 63 = 337$.

Answer: 337

Problem 8 Solution

Note $\triangle BAD$ has area 25, with two sides of $5\sqrt{2}$. Hence it is a 45-45-90 triangle with $BD = 10$ and BD is a diameter.

Let $BC = x$ and $CD = y$. Then $\triangle BCD$ is a right triangle with $x^2 + y^2 = 10^2$ and area $\frac{1}{2}xy$. Using Ptolemy's theorem,

$$AC \cdot BD = AB \cdot CD + AD \cdot BC$$
$$3\sqrt{10} \cdot 10 = 5\sqrt{2} \cdot y + 5\sqrt{2} \cdot x$$
$$30\sqrt{10} = 5\sqrt{2}(x + y)$$

Therefore $x + y = 6\sqrt{5}$ so

$$(x + y)^2 = (6\sqrt{5})^2$$
$$x^2 + 2xy + y^2 = 180$$

Combining with $x^2 + y^2 = 100$ we have

$$2xy = 180 - 100 = 80 \Rightarrow xy = 40.$$

Thus the area of $\triangle BCD$ is $\frac{1}{2} \cdot 40 = 20$.

Answer: 20

Problem 9 Solution

For (x,y) to be a lattice point on the curve, $x^2 + 14$ must be divisible by 15, or $x^2 + 14 \equiv 0 \pmod{15}$. Thus

$$x^2 \equiv -14 \equiv 1 \pmod{15}.$$

Clearly $x \equiv \pm 1 \equiv 1, 14 \pmod{15}$ will work. 15 is not prime, so there may be other solutions. Checking $\pm 2, \pm 3, \ldots, \pm 7$ we also have $x \equiv \pm 4 \equiv 4, 11 \pmod{15}$ also work. This gives

$$x = 1, 4, 11, 14, 16, 19, 26, 29, \ldots$$

as values that work.

However, as $0 \le x, y \le 100$ with $y = \dfrac{x^2 + 14}{15}$ we in fact have

$$\frac{x^2 + 14}{15} \le 100$$
$$x^2 + 14 \le 1500$$
$$x^2 \le 1486$$

Hence continuing the pattern from above the last values of x that work are $x = 31$ and $x = 34$ (as $41^2 > 1486$). This gives $4 + 4 + 2 = 10$ lattice points in total that work.

Answer: 10

Problem 10 Solution

The graph of $y = ||x| - 6|$ is symmetric with respect to the y-axis. Thus $L = -U$ so we restrict our attention to $m > 0$. Consider the graph of $y = ||x| - 6|$:

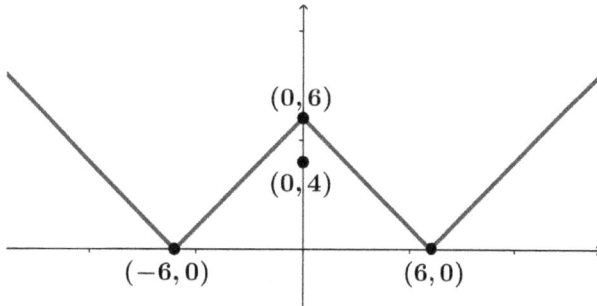

As long as $m < 1$ the graph will intersect the line $y = mx + 4$ twice when $x > 0$. To ensure the graphs intersect twice when $x < 0$, the line $y = mx + 4$ must have a positive y value when $x = -6$. Hence

$$-6m + 4 > 0 \Rightarrow m < \frac{2}{3}.$$

By the symmetry mentioned above, we must have $-\frac{2}{3} < m < \frac{2}{3}$ hence

$$U - L = \frac{2}{3} - \left(-\frac{2}{3}\right) = \frac{4}{3} \approx 1.3333.$$

Rounded to the nearest tenth our answer is 1.3.

Answer: 1.3

Problem 11 Solution

There are 3 base types of cupcakes, but as each can be bought with or without frosting, there are really 6 choices Patrick can choose from. Hence, using stars and bars, there are

$$\binom{8 + 6 - 1}{8} = 1287$$

ways Patrick can buy 8 cupcakes. However, there are

$$\binom{8+3-1}{8} = 45$$

outcomes where all 8 cupcakes are chosen without frosting. Therefore there are $1287 - 45 = 1242$ collections of cupcakes Patrick can buy.

Answer: 1242

Problem 12 Solution

An 5-digit palindrome must have the form $\overline{ABCBA}$ for digits A, B, and C. We want the palindrome to be divisible by 2, 3, 7, 11, and 13.

$7 \cdot 11 \cdot 13 = 1001$ so the number must be divisible by 1001. Thus $\overline{ABCBA} = 1001 \cdot K$ for some K. Note K must be a two-digit integer. However, then

$$1001 \cdot K = 1000K + K.$$

This has hundreds digit 0, so $C = 0$. Further, $\overline{AB} = \overline{BA}$ so $A = B$.

Thus the number needs to be of the form $\overline{AA0AA}$ and divisible by 6. A must be even, with $4A$ a multiple of 3, so $A = 6$. The palindrome is hence 66066.

Answer: 66066

Problem 13 Solution

Let G be the intersection of the medians $\overline{AD}$ and $\overline{BE}$ so G is the centroid.

As G is the centroid, $AG : DG = 2 : 1$, hence combining with $AD : AJ = 3 : 4$ the segments $DG = DJ$. Recalling D is the midpoint of $\overline{BC}$, the diagonals of $BGCJ$ bisect each other, so

$BGCJ$ is a parallelogram. An identical argument gives $AGCK$ is also a parallelogram.

For ease of calculation, let $[BDG] = 1$. This is one of the 6 triangles of equal area created by the medians of $\triangle ABC$, implying that $[ABC] = 6$. Diagonals of a parallelogram divide triangles into 4 equal pieces, thus, in particular,

$$[CDG] = [CDJ] = [CEG] = [CEK] = 1.$$

Lastly, as $BGCJ$ and $AGCK$ are parallelograms, so is $GKCJ$. Therefore

$$[CJK] = [GKCJ] \div 2 = 4 \div 2 = 2,$$

so the ratio $[ABC] : [CJK] = 6 : 2 = 3 : 1$ so $Q - P = 1 - 3 = -2$.

Answer: -2

Problem 14 Solution

We want the sum $1^3 + 2^3 + 3^3 \cdots + 2019^3$. Using the sum of two cubes formula $a^3 + b^3 = (a+b)(a^2 - ab + b^2)$ and grouping we have

$$2^3 + 2019^3 = (2 + 2019)(\cdots) \equiv 0 \pmod{2021}$$
$$3^3 + 2018^3 = (3 + 2018)(\cdots) \equiv 0 \pmod{2021}$$
$$\cdots$$

Therefore

$$1^3 + 2^3 + \cdots + 2019^3 \equiv 1^3 \pmod{2021}$$

so the remainder is 1.

Answer: 1

Problem 15 Solution
Let O be the center of the circle, and consider the triangles drawn below.

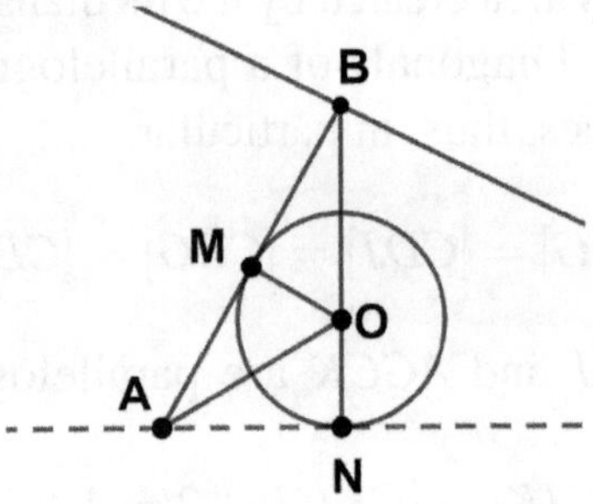

Right triangles $\triangle AMO$, $\triangle BMO$ and $\triangle ANO$ are all congruent. As $AB = 6$, $AM = 3$. The radius $MO = \sqrt{3}$, so $AO = 2\sqrt{3}$ and all three triangles are 30-60-90 triangles.

Hence $BN = \sqrt{3} + 2\sqrt{3} = 3\sqrt{3}$. $BC = CD \div 2 = 12$ so using the Pythagorean theorem

$$BC^2 = BN^2 + CN^2$$
$$12^2 = (3\sqrt{3})^2 + CN^2$$
$$144 = 27 + CN^2$$

Therefore, $CN^2 = 144 - 27 = 117$ so $CN = \sqrt{117}$ and $R = 117$.

Answer: 117

Problem 16 Solution
In order to obtain a multiple of 6 when she multiplies the numbers on the cards, she needs either (i) both cards being a multiple of 6, (ii) one card that is a multiple of 6 and any other card, or (iii) one card that is a multiple of 2 but not 3, and one card that is a multiple of 3 but not 2.

There are 2 cards that are multiples of 6: 6 and 12; there are 5 cards that are multiples of 2 but not 3; and there are 3 cards that are multiples of 3 but not 2.

Then, there is only 1 way to choose both cards multiples of 6; there are $2 \times 13 = 26$ ways to choose one multiple of 6 and any other card (that is not a multiple of 6 too); and there are $5 \times 3 = 15$ ways to choose one multiple of 2 and one multiple of 3 (both none a multiple of both).

Since there are $\binom{15}{2} = 105$ ways to choose two cards, the probability that she picks two cards whose product is a multiple of 6 is $\dfrac{1+26+15}{105} = \dfrac{42}{105} = \dfrac{2}{5}$. Therefore $Q - P = 5 - 2 = 3$.

Answer: 3

Problem 17 Solution

We first calculate the area of $A''E''F''$. Using $[-]$ to denote area, we calculate the area of $[AEF]$ indirectly as

$$[AEF] = [ABCD] - [ABE] - [ADF] - [CEF]$$
$$= 1 - \frac{1}{4} - \frac{1}{4} - \frac{1}{8}$$
$$= \frac{3}{8}$$

Since $\triangle A''E''F'' \sim \triangle AEF$ with a side length ratio of $3 : 1$,

$$[A''E''F''] = 3^2 \cdot [AEF] = \frac{27}{8}.$$

As $E'' = (3, 1.5)$ and $F'' = (1.5, 3)$ the area inside the square is as shown below

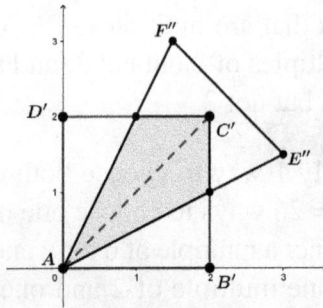

The quadrilateral inside $A'B'C'D'$ can be broken into two congruent triangles, each with base 1 and height 2 (as the sides of $A''E''F''$ intersect the midpoints of $\overline{B'C'}$ and $\overline{C'D'}$). Hence this shaded region has area $1 + 1 = 2$. Hence the fraction of $A''E''F''$ inside the square is

$$2 \div \frac{27}{8} = \frac{16}{27}$$

and $Q - P = 27 - 16 = 11$.

Answer: 11

Problem 18 Solution

Using Euler's totient function,

$$\phi(100) = 100 \cdot \frac{1}{2} \cdot \frac{1}{5} = 40$$

so as $\gcd(23, 100) = 1$, $23^{40} \equiv 1 \pmod{100}$. Since 2320 is a multiple of 40,

$$23^{2323} \equiv 23^3 \pmod{100}$$
$$\equiv 23^2 \cdot 23 \pmod{100}$$
$$\equiv 29 \cdot 23 \pmod{100}$$
$$\equiv 67 \pmod{100}.$$

Hence the last two digits of 23^{2323} are 67.

Answer: 67

Problem 19 Solution

Lines for $M = 1, 2$, and 3 are shown in reference to the square below.

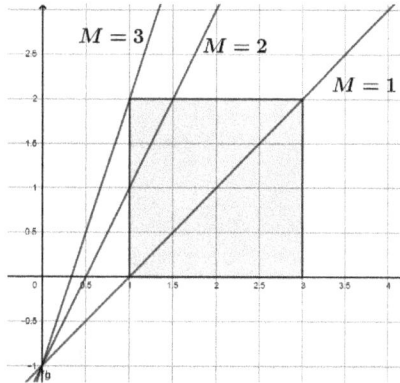

Note that for $M = 3, 4$, or 5 there is probability 0 that Polly's point is below the line. For $M = 2$, there is a $\dfrac{0.25}{4} = \dfrac{1}{16}$ chance of the point being above the line. For $M = 1$ there is a $\dfrac{1}{2}$ chance. Since each slope has $\dfrac{1}{5}$ chance of being chosen, the total probability is

$$\frac{1}{5} \cdot \frac{1}{2} + \frac{1}{5} \cdot \frac{1}{16}$$
$$= \frac{1}{10} + \frac{1}{80}$$
$$= 10\% + 1.25\%$$
$$= 11.25\%$$

so $K = 11.25$.

Answer: 11.25

Problem 20 Solution

We can write

$$\frac{r^2 + s^2 + t^2}{r + s + t} = \frac{(r+s+t)^2 - 2(rs + rt + st)}{r + s + t}.$$

By Vieta's theorem,

$$r + s + t = -\frac{-1}{2} = \frac{1}{2}$$

$$rs + rt + st = \frac{-13}{2}.$$

Hence the expression is

$$\left[\left(\frac{1}{2}\right)^2 - 2 \cdot \frac{-13}{2}\right] \div \frac{1}{2}$$

$$= 2 \cdot \left[\frac{1}{4} + 13\right]$$

$$= \frac{1}{2} + 26$$

$$= \frac{53}{2}.$$

Therefore $P + Q = 53 + 2 = 55$.

Answer: 55

2.8 ZIML May 2019 Junior Varsity

Below are the solutions from the Junior Varsity ZIML Competition held in May 2019.

The problems from the contest are available on p.71.

Problem 1 Solution

We want the remainder when 2^{503} is divided by 30, so work modulo 30. As $2^5 \equiv 2^1 \pmod{30}$, the pattern in the powers of 2^1, 2^2, 2^3, etc. will repeat every 4 terms. As 500 is a multiple of 4,

$$2^{503} \equiv 2^3 \equiv 8 \pmod{30},$$

so the remainder when 2^{503} is divided by 30 is 8.

Answer: 8

Problem 2 Solution

The polynomial $p(x)$ is already factored into 5 quadratics. Hence for $p(x)$ to have 10 distinct zeros, each of the 5 quadratics must have two distinct real roots.

Consider the first quadratic $x^2 + 6x + m$. Its discriminant is

$$6^2 - 4 \cdot m = 36 - 4m.$$

Thus this has two distinct real roots as long as $m < 9$. Since m is a positive integer, $m = 1, 2, \ldots, 8$.

All the other quadratics are identical to the above, except for an increasing linear term. Therefore each of their discriminant is larger than the discriminant of the first quadratic $x^2 + 6x + m$. Consequently, if the first quadratic has two real roots, all of the other quadratics do as well.

Hence $p(x)$ has 10 real roots for $m = 1, 2, \ldots, 8$, a total of 8 values of m.

Answer: 9

Problem 3 Solution

Let $BD = x$, so $CD = 26 - x$. Using the angle bisector theorem,

$$\frac{BD}{CD} = \frac{AB}{AC}$$
$$\frac{x}{26-x} = \frac{15}{37}$$
$$37x = 390 - 15x$$
$$52x = 390$$
$$x = 7.5$$

Thus $BD = 7.5$ with $BD^2 = 56.25$ and $CD = 18.5$ with $CD^2 = 342.25$.

Using Heron's formula for the area of $\triangle ABC$ we have

$$\sqrt{39 \cdot 24 \cdot 13 \cdot 2} = 156.$$

So the area of the heptagon is

$$56.25 + 342.25 + 156 = 554.5.$$

Answer: 554.5

Problem 4 Solution

The number's for Pete's pin must be in increasing order, so if 5 distinct numbers are chosen from 0 to 9 there is only one way to order them. There are $\binom{10}{5} = 252$ such ways.

These 252 pins include the cases where all the 5 digits are consecutive. Hence we need to subtract off the pins

$$01234, 12345, \ldots, 56789,$$

a total of 6. This gives an answer of $252 - 6 = 246$.

Answer: 246

Problem 5 Solution

$\overline{EB}$ is tangent to the circle at B, so is perpendicular to diameters $\overline{AB}$ and $\overline{CD}$ as shown in the diagram below.

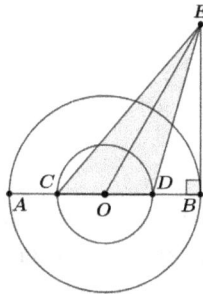

$OB = 2$ (the radius of the larger circle) and $EO = 4$, hence $\triangle EBO$ is a 30-60-90 triangle, with $EB = 2\sqrt{3}$. As this is also the height of $\triangle CDE$ (with base $CD = 2$, the diameter of the smaller circle), $\triangle CDE$ has area

$$\frac{1}{2} \cdot 2 \cdot 2\sqrt{3} = 2\sqrt{3}$$

so $R + S = 2 + 3 = 5$.

Answer: 5

Problem 6 Solution

Let $z = \sqrt{x-4}$, so $z^2 = x - 4$ and

$$x - 4\sqrt{x-4} - 13 = z^2 - 4z - 9.$$

Considering the quadratic function $g(z) = z^2 - 4z - 9$, this has a minimum when $z = \frac{-4}{-2} = 2$, so the minimum value is

$$g(2) = 2^2 - 4 \cdot 2 - 9 = -13.$$

Hence $f(x)$ also has a minimum value of -13, which occurs when $z = 2$. Therefore

$$\sqrt{x-4} = 2 \Rightarrow x = 4 + 4 = 8.$$

This gives $L + M = 8 + -13 = -5$ as the answer.

Answer: -5

Problem 7 Solution

Any number divisible by 9 has a sum of its digits that is also divisible by 9. Therefore all our sums will need to be divisible by 9.

It is impossible to have a 10-digit integer with sum of digits 0, so 9 is the smallest possible sum, achievable with the number $9,000,000,000$. The largest sum of digits is $10 \cdot 9 = 90$, achievable with the number $9,999,999,999$. Using only the digits 9 and 0, it is clear that all the sums

$$9, 18, 27, \ldots, 90$$

are possible, so there are 10 in total.

Answer: 10

Problem 8 Solution

The angles in a convex pentagon add up to $180° \cdot (5 - 2) = 540°$, so

$$\angle E = 540° - 4 \cdot 120° = 60°.$$

Since $120° = 2 \cdot 60°$, we can divide $ABCDE$ into 4 equilateral triangles, as shown below.

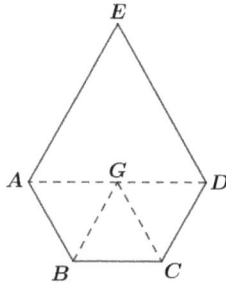

The combined area of the three congruent equilateral triangles $\triangle ABG$, $\triangle BCG$, and $\triangle CDG$ is equal to the area of $ABCD$, which is $27\sqrt{3}$, so each of the smaller equilateral triangles has area $9\sqrt{3}$.

Equilateral triangle $\triangle ADE$ has twice the side length, so has area $4 \cdot 9\sqrt{3} = 36\sqrt{3}$. Thus pentagon $ABCDE$ has area

$$27\sqrt{3} + 36\sqrt{3} = 63\sqrt{3}$$

and $K = 63$.

Answer: 63

Problem 9 Solution
Using Vieta's theorem we have

$$p + q = -4 \text{ and } r + s = 20.$$

If t is the common difference in the arithmetic sequence, then

$$(r + s) - (p + q) = 20 - (-4)$$
$$(2t + 3t) - (0t + 1t) = 24$$
$$4t = 24$$
$$t = 6$$

Therefore

$$r + s = 20$$
$$r + (r + 6) = 20$$
$$r = 7$$

and thus $s = 7 + 6 = 13$. Hence

$$D = r \cdot s = 7 \cdot 13 = 91$$

using Vieta's theorem once more.

Answer: 91

Problem 10 Solution

Sam's triangle (labeled $\triangle ABC$) and Jon's triangle (labeled $\triangle ADE$) are shown below.

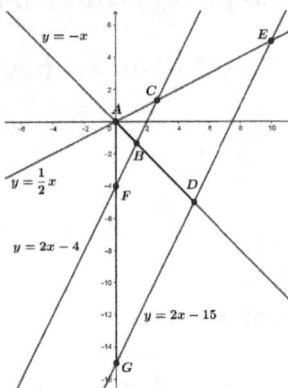

Note that since $y = 2x - 4$ and $y = 2x - 15$ are parallel, $\triangle ABC$ is similar to $\triangle ADE$. Let k denote the ratio of sides of $\triangle ABC$ to $\triangle ADE$. If $\triangle ABC$ has perimeter u and area v, then $\triangle ADE$ has perimeter ku and area k^2v. Hence

$$\frac{s}{t} = \left(\frac{v}{u}\right) \div \left(\frac{k^2v}{ku}\right) = \frac{1}{k}.$$

Again using that $y = 2x - 4$ and $y = 2x - 15$ are parallel, $\triangle ABF$ is similar to $\triangle ADG$. Therefore

$$k = \frac{AB}{AD} = \frac{AF}{AG} = \frac{4}{15},$$

so the ratio of $\dfrac{s}{t} = \dfrac{15}{4}$ with $P + Q = 15 + 4 = 19$.

Answer: 19

Problem 11 Solution

We are choosing R from an interval with total length 150.

Hence R can be closest to the odd perfect squares 1, 9, 25, up to 121. Consider the odd perfect square $(2k+1)^2$ and the two closest even perfect square $(2k)^2$ and $(2k+2)^2$. A real number r is closest to the odd perfect square if

$$\frac{(2k)^2 + (2k+1)^2}{2} < r < \frac{(2k+1)^2 + (2k+2)^2}{2}$$

$$\frac{8k^2 + 4k + 1}{2} < r < \frac{8k^2 + 12k + 5}{2}.$$

We want to find the combined lengths of these intervals for $k = 0$ to $k = 5$. Note this interval has length

$$\frac{8k^2 + 12k + 5}{2} - \frac{8k^2 + 4k + 1}{2}$$
$$= \frac{8k + 4}{2}$$
$$= 4k + 2.$$

Hence the length of the intervals closest to an odd perfect square are

$$2 + 6 + 10 + 14 + 18 + 22 = 72.$$

This gives a probability of

$$\frac{72}{150} = \frac{12}{25} = 48\%,$$

so $K = 48$.

Answer: 48

Problem 12 Solution

$\triangle ABC$ is a right triangle, hence the lengths of its sides form a Pythagorean triple. Therefore it's side lengths can be written as $k(m^2 - n^2)$, $k(2mn)$, and $k(m^2 + n^2)$ for relatively prime integers m and n and a constant integer k. The area of such a triangle is

$$\frac{1}{2} \cdot k(m^2 - n^2) \cdot k(2mn) = k^2 \cdot mn(m^2 - n^2).$$

With this, it is straightforward to make a table of possible areas for the area. Remember when looking at values of m and n we also need to consider scalings of the triangle (which multiply the area by a factor of 4, 9, 16, etc.). A table of all such values with area less than 100 is shown below.

m	n	$mn(m^2 - n^2)$	$\times 4$	$\times 9$	$\times 16$
2	1	6	24	54	96
3	1	24	96	–	–
4	1	60	–	–	–
3	2	30	–	–	–
4	3	84	–	–	–

From this table we see that the largest possible area is 96. (The triangle has sides 12, 16, and 20.)

Answer: 96

Problem 13 Solution

Since the sum of the 3 numbers is even, either the 3 numbers are all even, or one is even and the other two are odd. There are $\binom{5}{3} = 10$ ways to choose 3 even numbers, and there are $\binom{5}{1}\binom{5}{2} = 50$ ways to choose one even and two odd numbers. Thus there are 60 ways to choose 3 numbers whose sum is even.

Among the groups of 3 numbers with even sums, the following have their sum less than 10: $(0,2,4)$, $(0,2,6)$, $(0,1,3)$, $(0,1,5)$, $(0,1,7)$, $(0,3,5)$, $(2,1,3)$, $(2,1,5)$, $(4,1,3)$. There are 9 of these.

Therefore there are $60 - 9 = 51$ ways to choose 3 numbers so that their sum is even and at least 10.

Answer: 51

Problem 14 Solution
We know $x^3 + 5x - 3 = 0$ so $x^3 + 5x = 3$ as well as $x^3 = -5x + 3$. We use these facts to simplify $x^5 + 8x^3 - 3x^2 + 15x$, starting with any terms of degree ≥ 3:

$$
\begin{aligned}
&x^5 + 8x^3 - 3x^2 + 15x \\
&= x^3(x^2 + 8) - 3x^2 + 15x \\
&= (-5x + 3)(x^2 + 8) - 3x^2 + 15x \\
&= -5x^3 + 3x^2 - 40x + 24 - 3x^2 + 15x \\
&= -5x^3 - 25x + 24 \\
&= -5(x^3 + 5x) + 24
\end{aligned}
$$

Therefore, since $x^3 + 5x = 3$ the expression is equal to

$$-5 \cdot 3 + 24 = -15 + 24 = 9.$$

Answer: 9

Problem 15 Solution
First note that if $N > 100$ then $\gcd(N, 100) = \gcd(N - 100, 100)$ using the Euclidean algorithms. Thus it suffices to count N from 1 to 100 and multiply the answer by 10.

Recall Euler's totient (or phi) function $\phi(x)$ counts the number of integers from 1 to x that are relatively prime to x. Since

$\phi(100) = 40$, there are

$$100 - \phi(100) = 100 - 40 = 60$$

integers N from 1 to 100 with $\gcd(N, 100) > 1$. Therefore, there are 600 integers N from 1 to 1000 satisfy $\gcd(N, 100) > 1$.

Answer: 600

Problem 16 Solution
The question asks how many arrangements are possible. Regardless of how Pauli actually places the cards, at the end we have 5 black cards, each separated by at least one red card. Since the black cards are identical, count the number of ways to place the 7 different red cards into 4 spots (in-between the black cards).

Ignoring that the red cards are different, we use the positive version of stars and bars to get there are

$$\binom{7-1}{4-1} = \binom{6}{3} = 20$$

ways to determine how many red cards are in each of the 4 spots. Then there are 7! ways to order the 7 red cards. This gives

$$20 \cdot 7! = 20 \cdot 5040 = 100800$$

different arrangements.

Answer: 100800

Problem 17 Solution
Let r, s, and t be the roots of the equation.

Let $z = 2^x$, so our equation becomes $z^3 - 16z^2 + 60z - 64 = 0$. Using Vieta's theorem, the sum of the roots to this equation is $-(-16) = 16$ and the product of the roots is $-(-64) = 64$.

However, these are the roots after the substitution, so actually say

$$2^r + 2^s + 2^t = 16$$
$$\text{and } 2^r \cdot 2^s \cdot 2^t = 64$$

Note the second of these implies that $2^{r+s+t} = 64$, so as $64 = 2^6$, giving $r + s + t = 6$ as the answer.

Answer: 6

Problem 18 Solution

Let $N = \overline{ABCDE}$ where A, B, C, etc. are chosen from 0, 2, 3, etc.

Since N is divisible by $22 = 2 \cdot 11$, it is divisible by 11, hence the alternating sum of the digits $A - B + C - D + E$ is divisible by 11. Note this is only possible if

$$\{A, C, E\} = \{2, 3, 4\} \text{ and } \{B, D\} = \{0, 9\}.$$

Since N is a perfect square, it is divisible by $2^2 = 4$, so the last two digits are divisible by 4. Combining this with the above,

$$\overline{DE} = 04 \text{ or } 92.$$

Looking again at the digits, note $0 + 2 + 3 + 4 + 9 = 18$, so N is divisible by 9 as well. Therefore N must be a multiple of

$$2^2 \cdot 9 \cdot 11^2 = 4356.$$

As N is a 5-digit perfect square, N is one of

$$4356 \cdot 4, 4356 \cdot 9, \text{ or } 4356 \cdot 16.$$

Only $4356 \cdot 9 = 39204$ matches the last two digits, so $N = 39204$. (Double checking $N = 2^2 \cdot 3^4 \cdot 11^2$.)

Answer: 39204

Problem 19 Solution

The sphere has radius 1, so any circle on the sphere containing the center has circumference of 2π. Therefore an arc of length $\dfrac{\pi}{2}$ on the sphere is a quarter circle, while an arc of length $\dfrac{\pi}{4}$ is an eighth of a circle. As $\overset{\frown}{AB}$ and $\overset{\frown}{AC}$ meet $\overset{\frown}{BC}$ at right angles, viewing A as a pole of a sphere gives the diagram below.

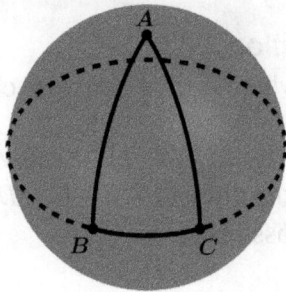

The full sphere has surface area 4π. Since $\overset{\frown}{AB}$ and $\overset{\frown}{AC}$ are quarter circles, the dotted circle containing $\overset{\frown}{BC}$ divides the surface area of the sphere in half. Further, since $\overset{\frown}{BC}$ is an eighth of a circle, the surface area of the region between the three arcs is

$$\frac{1}{2} \cdot \frac{1}{8} \cdot 4\pi = \frac{\pi}{4},$$

and thus $m = \dfrac{1}{4} = 0.25$.

Answer: 0.25

Problem 20 Solution

Erin gets to pick first with probability $\dfrac{5}{100} = \dfrac{1}{20}$.

Erin gets to pick second if Fran, Greg, Helen, Iris, or John pick first and then she picks second.

Both Fran and Greg pick first with probability $\dfrac{10}{100} = \dfrac{1}{10}$, then Erin gets to pick second with probability $\dfrac{5}{90} = \dfrac{1}{18}$. Thus Erin picks first with probability

$$\frac{1}{10} \cdot \frac{1}{18} = \frac{1}{180},$$

in these two cases.

All of Helen, Iris, and John pick first with probability $\dfrac{25}{100} = \dfrac{1}{4}$, then Erin gets to pick second with probability $\dfrac{5}{75} = \dfrac{1}{15}$. Thus Erin picks first with probability

$$\frac{1}{4} \cdot \frac{1}{15} = \frac{1}{60},$$

in these three cases.

Combining, the probability that Erin picks first or second is

$$\frac{1}{20} + 2 \cdot \frac{1}{180} + 3 \cdot \frac{1}{60} = \frac{1}{9},$$

and $Q - P = 9 - 1 = 8$.

Answer: 8

2.9 ZIML June 2019 Junior Varsity

Below are the solutions from the Junior Varsity ZIML Competition held in June 2019.

The problems from the contest are available on p.77.

Problem 1 Solution

Factoring inside each square root we have

$$-x^2 - 3x + 10 = -(x-2)(x+5)$$
$$\text{and } -x^2 + x + 6 = -(x-3)(x+2)$$

Both these quadratics have a negative leading coefficient, so they are positive in-between their roots. Hence for $f(x)$ to be defined we need $-5 \le x \le 2$ (from the first quadratic) and $-2 \le x \le 3$ (from the second quadratic. Therefore the domain of $f(x)$ is all x with $-2 \le x \le 2$.

We also need $f(x)$ to be positive. Note $f(x) \ge 0$ if and only if the first quadratic is greater than or equal to the second quadratic. Hence $f(x) \ge 0$ if and only if

$$-x^2 - 3x + 10 \ge -x^2 + x + 6$$
$$-4x \ge -4$$
$$x \le 1.$$

Thus for $f(x)$ to be defined and non-negative, $-2 \le x \le 2$ and $x \le 1$, therefore x must be in the interval $-2 \le x \le 1$ and $S = [-2, 1]$ with $L \cdot U = -2 \cdot 1 = -2$.

Answer: -2

Problem 2 Solution

There are $3 + 1 = 4$ middle school members and $4 + 2 = 6$ high school members. Since two middle school members and two high school members are chosen at random, there are

$$\binom{4}{2} \cdot \binom{6}{2} = 6 \cdot 15 = 90$$

total outcomes.

For 3 girls and 1 boy to be chosen, consider two cases, based on whether the boy is in middle school or in high school. This gives

$$\binom{1}{1} \cdot \binom{3}{1} \cdot \binom{4}{0} \cdot \binom{2}{2} + \binom{1}{0} \cdot \binom{3}{2} \cdot \binom{4}{1} \cdot \binom{2}{1}$$
$$= 1 \cdot 3 \cdot 1 \cdot 1 + 1 \cdot 3 \cdot 4 \cdot 2$$
$$= 27$$

outcomes. This gives a probability of

$$\frac{27}{90} = \frac{3}{10}$$

with $P + Q = 3 + 10 = 13$.

Answer: 13

Problem 3 Solution

Recognizing that $(16, 30, 34)$ is twice the Pythagorean triple $(8, 15, 17)$ the triangle is right. Combined with the fact that the center of the inscribed circle is the intersection of the angle bisectors we have the diagram

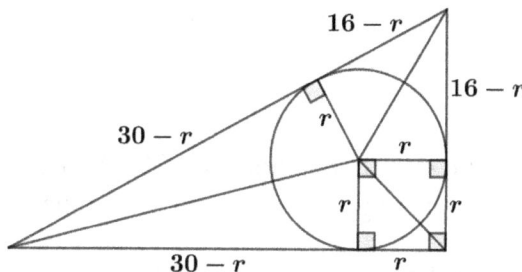

Therefore,

$$(30 - r) + (16 - r) = 34$$
$$46 - 2r = 34$$
$$r = 6.$$

Thus the area of the circle is $\pi \cdot 6^2 = 36\pi$ and $K = 36$.

Answer: 36

Problem 4 Solution

We want the remainder when a_{1771} is divided by 17, so we work modulo 17.

$a_0 = 5$ so using the recurrence $a_{n+1} = 4 \cdot a_n + 3$ we have

$$a_1 = 4 \cdot 5 + 3 \equiv 6 \pmod{17}$$
$$a_2 = 4 \cdot 6 + 3 \equiv 10 \pmod{17}$$
$$a_3 = 4 \cdot 10 + 3 \equiv 9 \pmod{17}$$
$$a_4 = 4 \cdot 9 + 3 \equiv 5 \pmod{17},$$

so $a_0 = a_4$. This also implies $a_{n+4} = a_n$ for all $n \geq 0$, so the pattern repeats every four terms. $1771 \div 4$ has remainder 3, so a_{1771} has remainder 9 when divided by 17.

Answer: 9

Problem 5 Solution

Peter and Paul arrive between 8am and 11am, a three hour time interval. Graph Paul's arrival time on the x-axis with Peter's on the y-axis. Breaking the three hours into 6 half-hour intervals, both their arrivals can be graphed in a 6×6 grid as the point (x, y).

Paul works out exactly one hour, so if Paul arrives first at time x, he will still be at the gym if Peter arrives at time y with $y < x + 1$.

If Peter arrives between 8am and 8:30am, he will work out until

9am (since he will then have worked out at least 30 minutes when it is 9am). If he arrives between 8:30am and 9:30am, he will work out until 10am. If he arrives after 9:30am he'll work out until at least 11am.

Combining both situations gives the graph below, where the shaded region represents the arrival times where Peter and Paul are both at the gym at the same time.

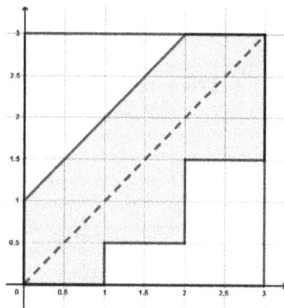

The 3×3 grid has area 9. The shaded region is the entire square minus a 2×2 triangle, a 1×0.5 rectangle, and a 1×1.5 rectangle. Hence the probability is

$$\frac{9 - 2 - 0.5 - 1.5}{9} = \frac{5}{9} = 55.\overline{5}\%.$$

Therefore, $K = 56$ when rounded to the nearest integer.

Answer: 56

Problem 6 Solution

$\triangle ABD$ is a right triangle, hence using the Pythagorean theorem we have

$$
\begin{aligned}
AB &= \sqrt{AD^2 + BD^2} \\
&= \sqrt{12^2 + 18^2} \\
&= \sqrt{6^2(2^2 + 3^2)} \\
&= 6\sqrt{13}.
\end{aligned}
$$

As $\overline{AB} \parallel \overline{CD}$, $\triangle ABE \sim \triangle CDE$ with ratio of sides

$$
\frac{AB}{CD} = \frac{AE}{CE} = \frac{13}{5}.
$$

Therefore

$$
\begin{aligned}
CD &= \frac{5}{13} AB \\
&= \frac{5}{13} \cdot 6\sqrt{13} \\
&= \frac{30\sqrt{13}}{13}
\end{aligned}
$$

and $r + s + t = 30 + 13 + 13 = 56$.

Answer: 56

Problem 7 Solution

Let $M = 10a + b$. Therefore

$$
M_{53} = 53a + b.
$$

Similarly

$$
53_M = 5 \cdot M + 3 = 50a + 5b + 3.
$$

Setting these equal gives

$$
\begin{aligned}
53a + b &= 50a + 5b + 3 \\
3a &= 4b + 3.
\end{aligned}
$$

Thus b is a multiple of 3. This gives possible pairs of digits

$$(a,b) = (1,0),(5,3), \text{ or } (9,6).$$

Since $M > 10$ and M cannot contain the digits 5 or 3, $M = 96$.

Answer: 10

Problem 8 Solution

$|x^2 - 2|$ is $x^2 - 2$ if $|x| \geq \sqrt{2}$ and $2 - x^2$ if $|x| < \sqrt{2}$.

Consider first $|x| < \sqrt{2}$. Setting the y values equal we have

$$2x + 3 = 2 - x^2$$
$$x^2 + 2x + 1 = 0$$
$$(x+1)^2 = 0$$

so $x = -1$, giving us 1 solution. Thus the second case must have 2 additional solutions.

If $|x| \geq \sqrt{2}$ we have

$$x^2 - 2 = 2x + 3$$
$$x^2 - 2x - 5 = 0,$$

so using the quadratic formula we get solutions of

$$x = \frac{2 \pm \sqrt{24}}{2} = 1 \pm \sqrt{6}.$$

(The problem implies that $1 - \sqrt{6} < -\sqrt{2}$ and $1 + \sqrt{6} > \sqrt{2}$, but this can be double checked.) Therefore

$$(1 - \sqrt{6})(1 + \sqrt{6})(-1) = -5 \cdot -1 = 5$$

is the product of the x-values of the three solutions.

Answer: 5

Problem 9 Solution

Diagonals $\overline{AC}$ are perpendicular, so they create four right triangles $\triangle ABE$, $\triangle BCE$, $\triangle CDE$, and $\triangle ADE$ as in the diagram below:

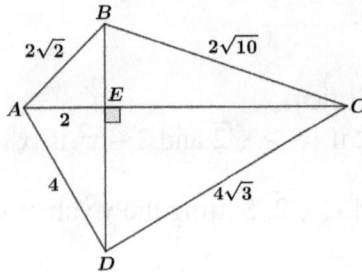

Recognizing $AD : AE = 2 : 1$, $\triangle ADE$ is a 30-60-90 triangle with $ED = 2\sqrt{3}$. Similarly $\triangle CDE$ is also a 30-60-90 triangle, with

$$CE = 2\sqrt{3} \cdot \sqrt{3} = 6.$$

Lastly, as $AB : AE = \sqrt{2} : 1$, $\triangle ABE$ is a 45-45-90 triangle so $BE = 2$.

Hence the diagonals of $ABCD$ have length $2 + 6 = 8$ and $2 + 2\sqrt{3}$, hence the area

$$S = \frac{1}{2} \cdot AC \cdot BD$$
$$= \frac{1}{2} \cdot 8 \cdot (2 + 2\sqrt{3})$$
$$= 8 + 8\sqrt{3}.$$

Using the fact that $\sqrt{3} \approx 1.73$,

$$S \approx 8 + 8 \cdot 1.73 = 21.84,$$

so rounded to the nearest integer the area is 22.

Answer: 22

Problem 10 Solution

The substitution $y = 2x^2 - 7x + 5$ gives

$$y^2 - 2y(10 - x) - 40x = 0$$
$$y^2 + (2x - 20)y - 40x = 0$$
$$(y - 20)(y + 2x) = 0.$$

(Here the factoring is done by considering the equation a quadratic in y.)

Hence either $y - 20 = 0$ or $y + 2x = 0$. In the first case

$$2x^2 - 7x - 15 = 0$$
$$(2x + 3)(x - 5) = 0$$

and $x = -\dfrac{3}{2}$ and $x = 5$ are solutions. In the second case

$$2x^2 - 5x + 5 = 0,$$

so there are no real solutions (the discriminant is $25 - 4 \cdot 2 \cdot 5 < 0$). Therefore the sum of the real solutions is

$$-\frac{3}{2} + 5 = \frac{7}{2} = 3.5,$$

giving an answer of 3.5.

Answer: 3.5

Problem 11 Solution

If the number $\overline{6a6b6c6d6}$ is divisible by 11, then its alternating sum of digits

$$6 - d + 6 - c + 6 - b + 6 - a + 6$$
$$= 30 - a - b - c - d$$
$$= 30 - (a + b + c + d)$$

is divisible by 11. a, b, c, and d are all digits, so

$$0 \leq a+b+c+d \leq 36$$

so the alternating sum must be one of 0, 11, or 22, implying that $a+b+c+d$ is one of

$$30-0 = 30, 30-11 = 19, \text{ or } 30-22 = 8.$$

To find the possible remainders when divided by 9, we need the sum of the digits

$$6+d+6+c+6+b+6+a+6$$
$$= a+b+c+d+30.$$

Using the possibilities for $a+b+c+d$, we have the sum of the digits is one of

$$30+30 = 60, 30+19 = 49, \text{ or } 30+8 = 38,$$

giving possible remainders of 6, 4, or 2. Thus the sum is $2+4+6 = 12$.

Answer: 12

Problem 12 Solution

We use complementary counting and subtract the number of ways where at least one of the cages has more than 4 birds.

Using stars and bars, there are

$$\binom{10+4-1}{10} = \frac{13}{10} = \frac{13 \cdot 12 \cdot 11}{6} = 286$$

total ways to place the birds into the cages.

To count the number of ways with at least one cage having more than 4 birds we use the principle of inclusion-exclusion. Let A, B,

C, and D denote the events that the first, second, third, and fourth cage (respectively) has more than 4 birds. Setting aside 5 birds (to ensure a cage has more than 4 birds) we have

$$n(A) = n(B) = n(C) = n(D)$$
$$= \binom{5+4-1}{5} = 56$$

(again using stars and bars). If two or more cages have more than 4 birds, each cage must have 5 birds. Hence

$$n(A \cap B) = n(A \cap C) = \cdots = n(C \cap D) = 1.$$

Combining we have

$$n(A \cup B \cup C \cup D)$$
$$= 4 \cdot 56 - 6 \cdot 1$$
$$= 224 - 6$$
$$= 218$$

This implies our final answer is $286 - 218 = 68$.

Answer: 68

Problem 13 Solution
Using only the difference of two squares, Brad factors the expression as

$$\left(x^{192} - y^{192}\right)$$
$$= \left(x^{96} + y^{96}\right)\left(x^{96} - y^{96}\right)$$
$$= \left(x^{96} + y^{96}\right)\left(x^{48} + y^{48}\right)\left(x^{48} - y^{48}\right)$$
$$= \cdots$$
$$= \left(x^{96} + y^{96}\right)\left(x^{48} + y^{48}\right)\left(x^{24} + y^{24}\right)\left(x^{12} + y^{12}\right)$$
$$\cdot \left(x^{6} + y^{6}\right)\left(x^{3} + y^{3}\right)\left(x^{3} - y^{3}\right)$$

Therefore $P = 7$. Using the difference of two squares, Claire factors

$$(x^{64} - y^{64})$$
$$= (x^{32} + y^{32})(x^{32} - y^{32})$$
$$= \cdots$$
$$= (x^{32} + y^{32})(x^{16} + y^{16})(x^8 + y^8)(x^4 + y^4)(x^2 + y^2)(x+y)(x-y)$$

Thus $Q = 7 + 1 = 8$. Hence $P + Q = 15$.

Answer: 15

Problem 14 Solution

Any factor of N has the form

$$2^a \cdot 3^b \cdot 5^c \cdot 7^d \cdot 11^e \cdot 13^f.$$

For the factor to end in 3 zeros, it must be divisible by $1000 = 2^3 \cdot 5^3$ so $a \geq 3$ and $c \geq 3$. For the factor to be a perfect cube, we must have all the exponents a multiple of 3. Therefore

$$a = 3 \text{ or } 6$$
$$b = 0, 3 \text{ or } 6$$
$$c = 3 \text{ or } 6$$
$$d = 0 \text{ or } 3$$
$$e = 0 \text{ or } 3$$
$$f = 0 \text{ or } 3.$$

This gives $2 \cdot 3 \cdot 2 \cdot 2^3 = 96$ possible factors.

Answer: 96

Problem 15 Solution

The arithmetic sequence has a common difference of $36 - 24 = 12$, so terms

$$24, 36, 48, 60, 72, 84, \text{ and } 96$$

that are less than 100, giving a total of 7 distinct red cards.

The geometric sequence has a common ratio of $36 \div 24 = 1.5$, so terms

$$24, 36, 54, \text{ and } 81$$

that are less than 100, giving a total of 4 distinct blue cards.

There are 7! ways to arrange the red cards, creating 8 spaces for the blue cards. Since the blue cards cannot be together, they can be placed in the spaces in $8 \cdot 7 \cdot 6 \cdot 5$ ways, and hence

$$N = 7! \cdot \frac{8!}{4!}.$$

This implies

$$
\begin{aligned}
N : 10! &= 7! \cdot 8 \cdot 7 \cdot 6 \cdot 5 : 10! \\
&= 7 \cdot 6 \cdot 5 : 10 \cdot 9 \\
&= 7 : 3
\end{aligned}
$$

with $P + Q = 7 + 3 = 10$.

Answer: 10

Problem 16 Solution
The sphere and cube share the same center, so by symmetry each edge of the cube is divided into portions of

$$\frac{1}{4} : \frac{1}{2} : \frac{1}{4} = 1 : 2 : 1$$

(with the middle portion inside the sphere).

In the diagram below, O is the center of the sphere and cube, C the middle of the bottom face of the cube, M is the midpoint of an edge, with the sphere intersecting the same edge at I and J.

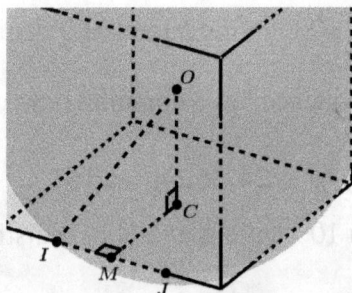

For convenience, assume the cube has side length 4, so $IJ = 2$ and $IM = 1$. Similarly, $CM = OC = 2$. Hence the radius

$$OI = \sqrt{OC^2 + CM^2 + IM^2}$$
$$= \sqrt{2^2 + 2^2 + 1^2}$$
$$= 3$$

using the distance formula.

Hence the sphere has volume

$$\frac{4}{3}\pi 3^3 = 36\pi.$$

Since the cube has volume $4^4 = 64$, the ratio of the volumes is

$$36\pi : 64 = 9\pi : 16$$

so $A + B = 9 + 16 = 25$.

Answer: 25

Problem 17 Solution

Consider the expression modulo 4, so we have

$$x^3 \equiv 9 \equiv 1 \pmod 4.$$

Therefore $x \equiv 1 \pmod 4$, so $x = 4k + 1$ for some k.

If $x = 4k + 1$ then

$$x^3 - 4y = 9$$
$$(4k + 1)^3 - 4y = 9$$
$$64k^3 + 48k^2 + 12k + 1 - 4y = 9$$
$$64k^3 + 48k^2 + 12k - 8 = 4y$$
$$16k^3 + 12k^2 + 3k - 2 = y.$$

Since $x \geq 0$, we must have $k \geq 0$. However, if $k = 0$, $y = -2$ which is not allowed. If $k = 1$, $x = 5$ and $y = 29$. If $k \geq 2$, $y > 100$, so in fact $(5, 29)$ is the only pair with $0 \leq x, y \leq 100$. Thus the answer is 1.

Answer: 1

Problem 18 Solution

If $\lfloor x \rfloor$ denotes the greatest integer $\leq x$, then for a positive integer k, there are $\lfloor \frac{1000}{k} \rfloor$ multiples of k among $1, 2, \ldots, 1000$. Hence there are

$$\lfloor \frac{1000}{6} \rfloor = 166 \text{ multiples of } 6$$
$$\lfloor \frac{1000}{10} \rfloor = 100 \text{ multiples of } 10$$
$$\lfloor \frac{1000}{15} \rfloor = 66 \text{ multiples of } 15$$

from 1 to 1000. However, any multiple of 30 is a multiple of all 3. (Further, a multiple of any 2 of the 3 numbers is also a multiple of the third.)

There are

$$\lfloor \frac{1000}{30} \rfloor = 33 \text{ multiples of } 30,$$

from 1 to 1000, which each of these multiples being triple counted (once for 6, 10, and 15). This gives a total of

$$166 + 100 + 66 - 3 \cdot 33 = 233$$

numbers from 1 to 1000 that are a multiple of 6, 10, or 15, but
not all 3.

Answer: 233

Problem 19 Solution

Using Vieta's theorem, the product of the four roots is $\dfrac{15625}{64}$ so

$$1 \cdot r \cdot r^2 \cdot r^3 = \frac{15625}{64}$$

$$r^6 = \frac{15625}{64}$$

$$r = \pm\frac{5}{2}.$$

Vieta's theorem also says the sum of the four roots is $-\dfrac{696}{64} < 0$

so $r < 0$. Thus $r = \dfrac{-5}{2}$ with $Q - P = 2 - (-5) = 7$.

Answer: 7

Problem 20 Solution

If $B = (x, y)$, then $y = x^2$ and $x^2 + y^2 = (2\sqrt{5})^2$. Hence

$$y + y^2 = 20$$

$$y^2 + y - 20 = 0$$

$$(y + 5)(y - 4) = 0$$

so $y = -5$ (impossible) or $y = 4$. Thus $B = (\sqrt{4}, 4)$ so $B = (2, 4)$
and the slope of $\overline{AB}$ is 2.

Therefore $\overline{BC}$ contains $(2, 4)$ and has slope $-\dfrac{1}{2}$, so is contained
in the line

$$y - 4 = -\frac{1}{2}(x - 2)$$

$$y = -\frac{1}{2}x + 5.$$

Solving for where it intersects the parabola we have

$$x^2 = -\frac{1}{2}x + 5$$
$$2x^2 + x - 10 = 0$$
$$(2x + 5)(x - 2) = 0$$

so $x = -\frac{5}{2}$ or $x = 2$. Hence $C = \left(-\frac{5}{2}, \frac{25}{4}\right)$ and

$$BC = \sqrt{\left(2 + \frac{5}{2}\right)^2 + \left(4 - \frac{25}{4}\right)^2}$$
$$= \sqrt{\frac{81}{4} + \frac{81}{16}}$$
$$= \sqrt{\frac{5 \cdot 81}{16}}$$
$$= \frac{9\sqrt{5}}{4}$$

Thus the area of $\triangle ABC$ is

$$\frac{1}{2} \cdot AB \cdot BC$$
$$= \frac{1}{2} \cdot 2\sqrt{5} \cdot \frac{9\sqrt{5}}{4}$$
$$= \frac{45}{4}$$
$$= 11.25$$

which gives the answer.

Answer: 11.25

3. Appendix

3.1 Junior Varsity Topics Covered

Algebra

- Students should be comfortable with ratios, proportions, and their applications to problems involving work and motion, but these problems are not a main focus at this level
- Exponents and Radicals: Laws of Exponents, Simplest Radical Form for Roots
- Factoring Tricks: Sums and differences of squares, cubes, etc., Binomial and multinomial theorem, Completing the Square/Rectangle, etc.
- Solving Equations: Linear Equations, Quadratic Equations, Systems of Equations, Substitutions to rewrite higher degree equations as quadratics, Radicals, Absolute Values
- Quadratics: Graphing and Vertex Form, Maxima and Minima, Quadratic Formula, Discriminant, Vieta's Theorem for sum and product of the roots

- Polynomials: Polynomial Long Division, Remainder and Factor Theorem, Rational Root Theorem, General Vieta's Theorem

Geometry

- As a general rule students should be comfortable using algebraic techniques (linear equations, quadratic equations, systems of equations, etc.) as tools for applying the geometric concepts listed below
- Angles in Parallel Lines (corresponding angles, alternating interior/exterior angles, same-side interior/exterior angles, etc.)
- Analytic Geometry: Equations of Lines, Parabolas, and Circles, Distance Formula, Midpoint Formula, Geometric Interpretation of Slope and Angles
- Triangles: Congruence and Similarity, Pythagorean theorem, Ratios of Sides for triangles with angles of 45, 45, 90 or 30, 60, 90
- Centers in Triangles: Definitions of altitudes, medians, angle bisectors, perpendicular bisectors, Definitions and basic properties of orthocenter, centroid, incenter, circumcenter, Angle Bisector Theorem
- Interior and Exterior Angles of Polygons, including the sum of all these angles, each angle if the polygon is regular, etc.
- Areas and Perimeters of basic shapes such as triangles, rectangles, parallelograms, trapezoids, and circles, Heron's formula and formulas using inradius or circumradius for triangles
- Geometric Reasoning with Areas: Congruent shapes have the same area, Similar triangles have a ratio of areas that is the square of the ratio of their sides, Triangles with the same height have a ratio of their areas equal to the ratio of their bases, etc., Using multiple expressions of area to solve for unknowns

- Circles: Arc Length, Sector Area, Definitions for Tangent Lines and Tangent Circles, Inscribed Angles, Angles formed by intersecting chords, Power of a Point, Ptolemy's Theorem
- Solid Geometry: Surface Area and Volume for Spheres, Prisms, Pyramids, and Cones, Reasoning for more general solids, such as combining the solids listed above or pieces of solids when cut by a plane, etc.

Counting and Probability

- Fundamental Rules: Sum and Product Rules, Permutations and Combinations
- Counting Methods: Complementary counting, Stars and bars (also called sticks and stones, balls and urns, etc.), Grouping objects that must be together, Inserting objects that must be apart into spaces between objects, etc., Principle of Inclusion and Exclusion
- Identities: Symmetry, Pascal's Identity, Hockey Stick Identity, etc. for binomial coefficients, Binomial and Multinomial Theorem, Understanding of these identities using combinatorial proofs
- Sequences: Arithmetic and Geometric Sequences and Series, Finding and understanding patterns and recursive definitions for general sequences
- Probability and Sets: Definitions for event, sample space, complement, intersection, and union, Understanding the use of Venn Diagrams
- Probability: In finite sample spaces as a ratio of the number of outcomes, In geometric sample spaces as a ratio of lengths, areas, or volumes, Axioms of Probability, Independence, Conditional Probability, Law of Total Probability

Number Theory

- Fundamental Definitions: Prime numbers, factors/divisors, multiples, least common multiple (LCM), greatest common factor/divisor (GCF or GCD), perfect squares/cubes/etc.
- Number Bases: Expressing and converting numbers in base 2, 3, 8, 16, etc, Understanding how to perform arithmetic in different bases
- Divisibility Rules for numbers such as 2, 3, 4, 5, 8, 9, 10, 11, and how to combine the rules for numbers such as 6, 22, etc.
- (Unique) Prime Factorization and how to use the prime factorization to find the number of factors, to test whether a number is a perfect square/cube/etc, to find the LCM or GCD.
- Factoring Tricks: Factors come in pairs, perfect squares have an odd number of factors, etc.
- Modular Arithmetic: Connection with remainders and applications such as "find the units digit", General rules for addition, subtraction, multiplication, and division, Extension of divisibility rules to calculating a number modulo 9, 11, etc., Fermat's Little Theorem, Euler's Totient Function and extension to Fermat's Little Theorem

3.2 Glossary of Common Math Terms

Acute Angle An angle less than $90°$.

Altitude of a Triangle A line segment connecting a vertex of a triangle to the opposite side forming a right angle. Also called the height of a triangle.

Angle A figure formed by two rays sharing a common vertex. Often measured in degrees.

Angle Bisector A line dividing an angle into two equal halves.

Arc The curve of a circle connecting two points.

Area The amount of space a region takes up. Often denoted using square brackets: area of $\triangle ABC = [ABC]$.

Arithmetic Sequence A sequence where the difference between one term and the next is constant.

Average See Mean.

Base of a Triangle One side of a triangle, often used when the altitude is drawn from the opposite side to this base.

Binomial Coefficient The symbol $\binom{n}{k} = \dfrac{n!}{k!(n-k)!}$.

Centroid of a Triangle The intersection of the three medians in a triangle.

Chord A line segment connecting two points on the outside of a circle.

Circle A round shape consisting of points that all have the same distance (called the radius) from the center of the circle.

Circumcenter of a Triangle The intersection of the three perpendicular bisectors in a triangle. Also the center of the circle that circumscribes a triangle.

Circumference The perimeter of a circle.

Circumscribe To draw a shape outside another shape so that the boundaries touch.

Coefficient The number being multiplied by a variable or power of a variable. For example, the coefficient of x^3 in $5x^5 + 4x^3 + 2x$ is 4.

Complement In probability, the complement of a set is all elements outside the set.

Composite Number A number that is not prime.

Congruent Two shapes or figures that are exactly the same.

Cube A solid figure formed by 6 congruent squares that all meet at right angles.

Deck of Cards A standard deck of cards has 52 cards. There are 4 suits (clubs, diamonds, hearts, and spades) with each suit having cards of 13 ranks (A (ace), $2, 3, \ldots, 10$, J (jack), Q (queen), and K (king)).

Degree of a Polynomial The highest power of a variable in the polynomial. For example, the degree of $2x^3 - 5x^6 + 2$ is 6.

Denominator The bottom number in a fraction.

Diagonal A line segment connecting two vertices of a shape or solid that is not an edge of the shape or solid.

Diameter A chord passing through the center of a circle. The diameter has length that is twice the radius.

Die or Dice A standard die (plural is dice) has 6 sides. Each of the 6 sides has the same chance when the die is rolled.

Digit One of $0, 1, 2, \ldots, 9$ used when writing a number.

Discriminant The expression $b^2 - 4ac$ for a quadratic equation $ax^2 + bx + c = 0$.

Distinguishable Objects Objects that are different.

Divisible A number is divisible by another number if there is no remainder when the first number is divided by the second. For example, 35 is divisible by 7.

Divisor A number that evenly divides another number. For example, 6 is a divisor of 48. Also called a factor.

Edge A line segment connecting two vertices on the outside of a shape or solid.

Equally Likely Having the same chance of occurring.

Equiangular Polygon A shape with all equal angles.

Equilateral Polygon A shape with all equal sides.

Equilateral Triangle A regular triangle, one with three equal sides and three equal angles.

Even Number A number divisible by 2.

Exponent The number another number is raised to for powers. For example, in a to the power of b (a^b), the exponent is b.

Face The shape or polygon on the outside of a solid region.

Factor of a Number A number that evenly divides another number. For example, 6 is a factor of 48. Also called a divisor.

Factorial The symbol ! where $n! = n \times (n-1) \times (n-2) \cdots \times 1$.

Fraction An expression of a quotient. For example, $\dfrac{1}{2}$ or $\dfrac{9}{7}$.

Function A function is a rule that associates exactly one output with every input. Often described using an equation.

Geometric Sequence A sequence where the ratio between one term and the next is constant.

Greatest Common Divisor/Factor (GCD/GCF) The largest number that is a divisor/factor of two or more numbers.

Incenter of a Triangle The intersection of the three angle bisectors in a triangle. Also the center of a circle that is inscribed inside a triangle.

Indistinguishable Objects Objects that are the same.

Inscribe To draw a shape inside another shape so that the boundaries touch.

Intersecting Lines or curves that cross each other.

Intersection of Two Sets The set of objects that are in both of the two sets. Denoted using $\cap$. For example, $\{2,3\} \cap \{3,4,5\} = \{3\}$.

Isosceles Triangle A triangle with two equal sides and two equal angles.

Least Common Multiple (LCM) The smallest number that is a multiple of two or more numbers.

Mean The sum of the numbers in a list divided by the how many numbers occur in the list. Also called the average.

Median The number in the middle of a list when the list is arranged in increasing order.

Median of a Triangle A line connecting a vertex in a triangle to the midpoint of the opposite side.

Midpoint The point in the middle of a line segment.

Mode The number or numbers occurring most often in a list of numbers.

Multiple A number that is an integer times another number. For example, 72 is a multiple of 8.

Numerator The top number in a fraction.

Obtuse Angle An angle between $90°$ and $180°$.

Odd Number A number not divisible by 2.

Orthocenter of a Triangle The intersection of the three altitudes in a triangle.

Parallel Lines Lines that do not intersect.

Perfect Cube A number that is another number cubed. For example, $64 = 4^3$ is a perfect cube.

Perfect Square A number that is another number squared. For example, $64 = 8^2$ is a perfect square.

Perimeter The length/distance around the outside of a shape.

Perpendicular Bisector A line perpendicular to and passing through the midpoint of a line segment.

Pi (π) A number used often in geometry. $\pi = 3.1415926\ldots \approx 3.14 \approx \dfrac{22}{7}$.

Polygon A shape formed by connected line segments.

Polynomial A function that is made of adding multiples of powers of a variable. For example, $f(x) = x^4 + 3x^2 + 2x - 3$.

Prime Factorization The expression of a number as the product of all its prime factors. For example, 24 has prime factorization $2 \times 2 \times 2 \times 3 = 2^3 \times 3$.

Prime Number A number whose only factors are one and itself.

Proportional Ratios Ratios that have equal values when expressed in fraction form. For example, $2 : 3$ is proportional to $8 : 12$.

Quadratic A polynomial with degree 2. Often written in the form $ax^2 + bx + c$.

Quadrilateral A shape with four sides.

Quotient The integer quantity when dividing one number by another. For example, the quotient of $38 \div 5$ is 7 as $38 = 7 \times 5 + 3$.

Radius of a Circle The distance from the center of the circle to any point on the outside of the circle.

Randomly Chosen for a group of objects. Unless specified, the chance of choosing each object is the same as any other object.

Rank of a Card See Deck of Cards.

Ratio A relation depicting the relation between two quantities. For example $2 : 3$ or $\frac{2}{3}$ denotes that for every 3 of the second quantity there are 2 of the first quantity.

Rational Number A number that can be written as a fraction.

Reciprocal One divided by the number. For example, the reciprocal of 7 is $\frac{1}{7}$.

Rectangle A quadrilateral with four right angles (an equiangular quadrilateral).

Regular Polygon A polygon with all equal sides and all equal angles (equilateral and equiangular).

Remainder The quantity left over when one integer is divided by another. For example, the remainder of $38 \div 5$ is 3 as $38 = 7 \times 5 + 3$.

Rhombus A quadrilateral with four equal sides (an equilateral quadrilateral).

Right Angle A $90°$ angle.

Right Triangle A triangle containing a right angle.

Root of a Function A value of x such that the function evaluates to zero. For example, $x = 2$ is a root of the function $f(x) = x^2 - 4$.

Sample Space In probability, the sample space is the set of all outcomes for an experiment.

Scalene Triangle A triangle with three unequal sides and three unequal angles.

Sector The region formed by an arc and the two radii connecting the ends of the arc to the center of the circle.

Sequence An ordered list of numbers.

Set An unordered collection or group of objects without repeated elements. Denoted using curly brackets. For example, $\{1,2,3,4\}$ is the set containing the integers $1,\ldots,4$.

Similar Shapes or solids that have the same angles and sides that share a common ratio.

Simplest Radical Form An expression containing a radical such that the number inside the radical is an integer that has no perfect squares.

Sphere A round solid consisting of points that all have the same distance (called the radius) from the center of the sphere.

Square A shape with four equal sides and four equal angles (a regular quadrilateral).

Subset A set of objects that is contained inside a larger set of objects. Denoted using $\subseteq$. For example $\{2,3\} \subseteq \{1,2,3,4\}$.

Suit of a Card See Deck of Cards.

Surface Area The total area of all the faces of a solid.

Tangent Line A line touching a shape or curve at exactly one point.

Trapezoid A quadrilateral with one pair of parallel sides.

Triangle A shape with three sides.

Union of Two Sets The set of objects that are in one or both of the two sets. Denoted using $\cup$. For example, $\{2,3\} \cup \{3,4,5\} = \{2,3,4,5\}$.

Venn Diagram A diagram with circles used to understand the relationship between overlapping sets.

Vertex The intersection of line segments, especially the intersection of sides or edges in a shape or solid.

Volume The amount of space a solid region takes up.

With Replacement When choosing objects with replacement, a chosen object is returned to the others allowing it to be chosen more than once.

3.3 ZIML Answers

ZIML October 2018 Junior Varsity

Problem 1:   −6   Problem 11:   55

Problem 2:   31104   Problem 12:   16

Problem 3:   175252   Problem 13:   11104

Problem 4:   130   Problem 14:   1.5

Problem 5:   17424   Problem 15:   5

Problem 6:   3   Problem 16:   199

Problem 7:   104   Problem 17:   −8

Problem 8:   70   Problem 18:   2.8

Problem 9:   4   Problem 19:   10

Problem 10:   120   Problem 20:   220

ZIML November 2018 Junior Varsity

Problem 1: 11001010011

Problem 2: 21600

Problem 3: 1

Problem 4: 2

Problem 5: 12

Problem 6: 5

Problem 7: 108

Problem 8: 7

Problem 9: -1.75

Problem 10: 23

Problem 11: -4

Problem 12: 19

Problem 13: 55

Problem 14: 5

Problem 15: 225

Problem 16: 40

Problem 17: 64

Problem 18: 1

Problem 19: 5

Problem 20: 46

ZIML December 2018 Junior Varsity

Problem 1:	1440	**Problem 11:**	5
Problem 2:	45	**Problem 12:**	20
Problem 3:	1122	**Problem 13:**	4
Problem 4:	3	**Problem 14:**	4
Problem 5:	585	**Problem 15:**	9
Problem 6:	14280	**Problem 16:**	37.5
Problem 7:	20	**Problem 17:**	-1
Problem 8:	135	**Problem 18:**	-12
Problem 9:	-8	**Problem 19:**	90
Problem 10:	122	**Problem 20:**	5

ZIML January 2019 Junior Varsity

Problem 1:	20	**Problem 11:**	1.75
Problem 2:	14159	**Problem 12:**	3.2
Problem 3:	1146	**Problem 13:**	1000
Problem 4:	330	**Problem 14:**	5
Problem 5:	15	**Problem 15:**	-6
Problem 6:	60	**Problem 16:**	119
Problem 7:	17	**Problem 17:**	10
Problem 8:	14851	**Problem 18:**	7
Problem 9:	48	**Problem 19:**	360
Problem 10:	2	**Problem 20:**	230

ZIML February 2019 Junior Varsity

Problem 1:	30	**Problem 11:**	2
Problem 2:	2020200	**Problem 12:**	16
Problem 3:	1680	**Problem 13:**	10
Problem 4:	-11	**Problem 14:**	-2496
Problem 5:	3.75	**Problem 15:**	200
Problem 6:	2501	**Problem 16:**	160000
Problem 7:	1.75	**Problem 17:**	37
Problem 8:	42	**Problem 18:**	150
Problem 9:	-36	**Problem 19:**	18
Problem 10:	10	**Problem 20:**	-4

ZIML March 2019 Junior Varsity

Problem 1:	84	Problem 11:	7
Problem 2:	−192	Problem 12:	1
Problem 3:	4	Problem 13:	210
Problem 4:	82	Problem 14:	16.8
Problem 5:	9	Problem 15:	9
Problem 6:	97	Problem 16:	3
Problem 7:	37	Problem 17:	24
Problem 8:	91	Problem 18:	4
Problem 9:	20	Problem 19:	233
Problem 10:	923	Problem 20:	36

ZIML April 2019 Junior Varsity

Problem 1:	70	**Problem 11:**	1242
Problem 2:	17	**Problem 12:**	66066
Problem 3:	48	**Problem 13:**	-2
Problem 4:	70	**Problem 14:**	1
Problem 5:	33	**Problem 15:**	117
Problem 6:	3	**Problem 16:**	3
Problem 7:	337	**Problem 17:**	11
Problem 8:	20	**Problem 18:**	67
Problem 9:	10	**Problem 19:**	11.25
Problem 10:	1.3	**Problem 20:**	55

ZIML May 2019 Junior Varsity

Problem 1:	8	**Problem 11:**	48
Problem 2:	9	**Problem 12:**	96
Problem 3:	554.5	**Problem 13:**	51
Problem 4:	246	**Problem 14:**	9
Problem 5:	5	**Problem 15:**	600
Problem 6:	-5	**Problem 16:**	100800
Problem 7:	10	**Problem 17:**	6
Problem 8:	63	**Problem 18:**	39204
Problem 9:	91	**Problem 19:**	0.25
Problem 10:	19	**Problem 20:**	8

ZIML June 2019 Junior Varsity

Problem 1: -2 Problem 11:   12

Problem 2:   13   Problem 12:   68

Problem 3:   36   Problem 13:   15

Problem 4:   9   Problem 14:   96

Problem 5:   56   Problem 15:   10

Problem 6:   56   Problem 16:   25

Problem 7:   10   Problem 17:   1

Problem 8:   5   Problem 18:   233

Problem 9:   22   Problem 19:   7

Problem 10:   3.5   Problem 20:   11.25